200,000,000 Years Beneath the Sea

Also by Peter Briggs

200,000,000 YEARS BENEATH THE SEA

by Peter Briggs

HOLT, RINEHART AND WINSTON

NEW YORK   CHICAGO   SAN FRANCISCO

Published simultaneously in Canada by Holt, Rinehart
and Winston of Canada, Limited.

ISBN: 0-03-085983-2

Library of Congress Catalog Card Number: 78-138885

First Edition

DESIGNED AT THE INKWELL STUDIO

Printed in the United States of America

ACKNOWLEDGMENTS / A "thank you" page is the one page that everyone not personally involved skips over. However, I do want to express my gratitude in print to Tony Pimm, Henry J. Smith, Mel Peterson, Terry Edgar, Jack Renirie, Sam Gerard, Ken Brunot, Bill Ryan, Dinah Forbes-Robertson Sheean, Sue Thompson, Fran Parker, Tom Wiley, Archie McLerran, Darrell Sims, James Hays, Joshua Tracey, Dr. Tommy Austin, Art Maxwell, Bruce Heezen, Captain Joe Clarke, and the cooks on the *Challenger*.

This book is really about what wonderful things men can do when they act intelligently and peacefully together.

PREFACE / The greatest voyage of earth discovery in this century began August 11, 1968.

By drilling into the sediments on the ocean floor that covers three-quarters of the earth, scientists aboard the ship of discovery, the *Glomar Challenger*, have produced very persuasive confirmation of a number of startling theories about the oceans and the continents. One of the most spectacular ideas that appears to have been proved by examining these deep sediments is that the North Atlantic is no more than 200 million years old and that North America and Europe were once joined together. Drilling has also shown that the South Atlantic may be only 150 million years old and that South America has been separating from Africa at the rate of about two inches a year ever since their initial rupture.

Work aboard the *Challenger* also seems to have proved that the Pacific is no older than the Atlantic and has been spreading out in the same way, but that the movement of North and South America to the west has destroyed half the evidence. The men working with *Challenger* material have produced powerful evidence that the continents of the world are in constant motion, that they are "drifting." These motions of continents, and the vast earth plates on which they rest, along with adjacent seas, are believed to have given rise to the world's great mountain chains, earthquakes, and volcanoes. Oil discoveries made in deep waters will have immediate economic repercussions and make it extremely urgent that an international agreement about the control of the deep sea floor be reached.

This is the story of the ship, the men, their ideas, their evidence, and their cruises. The journeys of the *Glomar Challenger* have already made every geography book, every text in geology, obsolete. The voyages are continuing; the research on the sediments cored, sometimes from more than 3,000 feet beneath the sea floor, will go on for years. New, earth-shattering ideas will doubtless emerge, but the record as of today is quite fascinating enough to justify a report.

PETER BRIGGS

New York
March, 1971

200,000,000 Years Beneath the Sea

1

"The greatest surprise that the geologic study of the oceans has provided in the last decade is the youth of the oceanic crust as compared with the age of the crust of the continents."

—Alfred G. Fischer and Bruce C. Heezen, Co-Chief Scientists, Leg VI of the *Glomar Challenger* cruises

ON NOVEMBER 24, 1968, a strange-looking ship approached the long, gleaming white sand beaches of Cape Verde, the most western point of all Africa. A few hours later it sailed into Dakar, the modern city built by the French that is now the capital of the new nation of Senegal. To experienced watchers along the quays, it looked like a self-propelled rig for drilling oil. To camel-riding Arabs just off the desert, and to Africans fresh from the bush country, this ship must have looked even more strange. The ships they may have seen in the occasional movie they watched certainly were not like this. The watchers on the quay who recognized modern ships were correct. Basically, the *Glomar Challenger* had been built around an oil rig. The ship had just crossed the Atlantic and, using the experience gained in the offshore oil fields of the United States, had paused nine times in the crossing to drill holes in the sediment of the deep ocean floor. Oil, however, was not the wealth it was seeking.

Soon after the ship docked, the crew, who had been at sea for seven weeks, hurried ashore to soak their parched senses in the sights and sounds and smells of a living city full of strange people. They had a drink at a bar reminiscent of Paris, took a cab for a glimpse of the broad avenues and the shining new buildings, perhaps look at a Mohammedan mosque, brush quickly by the African quarter, and then catch the first plane that would fly them back home. A new crew waited in Dakar to take possession of the *Glomar Challenger* and sail on its next long leg over the high seas.

Two young geologists, Dr. Mel Peterson and Dr. Terry Edgar of the Scripps Institution of Oceanography in California, were among the last to go ashore. They had been the scientific directors of this, the *Glomar Challenger's* second cruise. Their mood was mixed now that the voyage had ended. In dispatches to headquarters at La Jolla in California, with copies to the offices of the National Science Foundation in Washington, D.C., they had radioed news of their first successes. The samples of the ocean floor they took by coring produced the first physical evidence that the sea floor of the Atlantic Ocean was spreading. They were the first oceanographers to reach the basement rock of the Atlantic by drilling. The sediments near the basement produced some very interesting new ideas on how the ores of metal are created. Much more information would certainly be revealed when the cores had undergone detailed analysis.

Yet the saga of Leg II of the *Glomar Challenger* was not written entirely without trouble. The difficulties had led the scientists to ask for an extension of the allotted sea time. The British-born, but somewhat Americanized, Terry Edgar said, "Gosh, we're really getting the hang of this and now it's all over!" For Edgar and for Mel

Peterson the time had gone too swiftly. In the overall Deep Sea Drilling Project, Dr. Peterson was the chief scientist and Dr. Edgar was the coordinating staff geologist. The two men had been in charge of this second leg of the nine legs that were planned because, to some extent, it was a test run—a time for debugging. One of their jobs was to make certain things would go as smoothly as possible for the scientists who were to follow.

On Leg I, between Galveston and New York, the scientists were able to recover about 40 percent of the material that presumably had been forced into the core barrel. On the Atlantic crossing, the first cores were only surfacing about 33 percent of the whole potential. This low productivity disturbed Mel Peterson and, after worrying over the matter for some time, he conceived the idea of a very simple core-catcher that would hold all the material that was falling out of the thirty-foot core barrel as it was being brought to the surface. This ingenious small device—so simple it was surprising no one had ever thought of it before—consisted of little frayed fingers, cut into the bottom of the plastic core, that would bend upward when the sediment was being intruded and then slip back to close up the opening at the bottom as the sediment slid downward. This core-catcher was so effective that it would even hold a column of water. And the cost for the material was only one cent American. For a single penny, the recovery rate of the object, for which millions of dollars were being spent, had been more than doubled. Productivity reached a point of about 90 percent efficiency.

Although this technical success was indeed cause for elation, other problems remained. In the process of drilling, the scientists on the second leg of the *Challenger* expedition ran across an obstacle that was to plague many future voyages. At the relatively shallow depth of the first site off the American side of Bermuda, they encountered a very hard layer of rock called chert. This chert, a formation somewhat like flint or porcelain, wore out the diamond bits after just a few feet of drilling and it was impossible to get past it to the sediments underneath. Strategic drilling sites, if they were covered by chert, had to be abandoned without any scientific dividend, even though great amounts of time and effort had gone into their selection and into getting the ship to the chosen location. Other technical difficulties appeared during the seven weeks at sea so that, in spite of all they had accomplished, Peterson and Edgar—well aware of the $26,000 a day it cost to operate the *Glomar Challenger*—went ashore in Dakar not completely satisfied. Of course, it was too much to expect perfection in every phase of the operation of a ship that was in so many ways unique. Other men might have been satisfied simply by the ship's performance in heavy seas. It showed that by use of its two thruster propellers on the bow and two in the stern it could stay almost perfectly in position, without moving, while as much as four miles of drilling pipe dangled through the hole built for it amidships.

The two scientists spent a week in Dakar before flying home to California. They arranged for the 715 feet of deep-sea sediment in the plastic core barrels to be shipped back to the specially built core library of the Deep Sea Drilling Project (DSDP) on the campus of the Lamont-Doherty Geological Observatory in Palisades, New York. Once back in the United States, selected bits of the cores would be examined in many laboratories around the country and probably in foreign countries as well. While waiting in Dakar for a variety of equipment to be flown in—material they learned they needed while far out at sea—Peterson and Edgar sorted some of the masses of data that had been collected. All of it would be needed to write the book, hundreds of pages long, that would be published about Leg II. Peterson and Edgar also took this time to tell the scientists going on the next leg all they had learned about the very complex business of drilling into the past history of the oceans.

Dakar had been chosen as the eastern terminus of this *Glomar Challenger* trip because it had one of the best harbors in West Africa and very modern facilities. It was also the obvious choice geologically. This western tip of Africa had been discovered in 1444 by a Portuguese explorer, Alvara Fernanda, in the service of Prince Henry the Navigator. Fernanda had dared to challenge the folktale that men would be burned up by the sun's rays as they approached the Equator, and then sailed on to look for a sea passage to India. What he did find was a point of land that one modern theory states was connected with Florida, somewhere between Cape Kennedy and Palm Beach, more than 100 million years ago.

Although Leg II had a number of good scientific stories connected with it, what publicists found most newsworthy was that the drilling offered proof of the theory that the sea floor was spreading. When the *Glomar Challenger* had reached Dakar and sent back the news, the Washington *Star* condensed the results into a headline that read, "It's Farther to Europe, Study Finds." Peterson and Edgar had calculated that, over the course they took, the Atlantic was spreading apart at the rate of about two inches a year. By this calculation, Dakar was 86 feet farther away from Florida than it had been in 1444 when Fernanda discovered it.

Naturally, the expeditions of the *Challenger* did not spring full-blown from the imaginations of the two scientists who were in charge of that Atlantic crossing. The original thought has been traced back to an observation made by Sir Francis Bacon, in 1620. The ships of Magellan had circumnavigated the globe one hundred years before, and by Bacon's time the contours of lands bordering the Atlantic had been fairly accurately charted. Though other men had probably also noted the same things, it was the English philosopher-scientist who first drew attention to the remarkable fit between the coastlines of Africa and South America and suggested that perhaps they had once been joined together.

No one had seriously disputed the idea—which had simply lain more or less dormant—when, at the end of the nineteenth century, the Austrian geologist Eduard Suess saw similarities in the rock formations of all the lands in the southern hemisphere, and offered the idea that they had all been joined together in one massive continent. He named this hypothetical land mass Gondwanaland, for the Gondwana province in India, whose unusual and very ancient geology led Suess to believe that it had once been at the very heart of the old continent. To many scientists, Gondwanaland amounted to nothing more than the kind of sensational, but unfounded, speculation you read in Sunday newspaper supplements. But Gondwana itself was special; it had the only coal found in India and most of the iron and manganese ores, and its fossils resembled those discovered in Africa.

A few years later a German, Alfred Wegener, went a step further toward reconstructing the geography of long ago. On the basis of extensive evidence from geology and paleontology that showed conspicuous similarities between the continents, Wegener suggested that all the lands in the world had once been joined together in one great body, which he named Pangaea. Geologists, particularly in the United States, scorned Wegener's idea, insisting that he was mistaken because he could suggest no force capable of moving such immense masses of material. Wegener's theory also suffered from the disclosure that calculations he used to show the relative movement of Greenland away from Europe were mathematically incorrect.

Yet geologists, confronted with the fossils of animals that had undoubtedly lived in the sea but had been found a thousand or more miles inland, were perfectly willing to accept the idea that parts of North America had been inundated by shallow seas at least eight or nine times in the geologic past. They saw that sea levels could change, that continents could subside and rise again, but they honestly could not see how these continents could float around the globe. In the United States, only petroleum geologists, who looked for evidence of ancient seas in order to drill for oil in such regions, believed that continental drifting could be a respectable assumption. Much later, a very distinguished South African geologist, du Toit, found considerable evidence of continental drifting in Africa, in South America, and in India as well, but his carefully researched papers could not at first attract a wide audience in universities north of the Equator.

Good scientific training tends to produce conservative men. Wild ideas come easily to fertile imaginations, particularly when those imaginations go to work on a single explanation for the many awesome scars and gashes on the earth's ancient crust. Obviously, the lands have gone through extreme changes in the past. How else can the various shapes of coastlines, the high mountain ranges and deep depressions, the vast plains and deserts be understood? Volcanoes and earthquakes likewise need to be explained. In the face of all the colorful explanations given by myth-makers and dreamers, scientists have learned to be very skeptical, to demand proof of everything, and to insist that each claim made must be proved by repeated independent experiments. When an idea is proposed that defies not only common sense, but also man's dignity itself (such as the popular misconception that Darwin had said we are all descended from apes), then the burden of proof has to be massive indeed before a respectable scientist, and much less the general public, will accept it. Scientists found many convincing reasons for resisting the notion of drifting continents. But a revolution in geological thought was stirring. A threat to entrenched ideas was coming from an unexpected quarter. Men had begun to explore a new frontier—the sea, and in particular, the solid floor beneath it.

Two apparently unrelated developments, both occurring soon after the death of Queen Victoria, had little effect on the fixed positions of conservative geologists. But later, as we shall see, these observations became quite important in the drift controversy. Two Englishmen, Robert Scott and Ernest Shackleton, found beds of coal, thought to be the fossil residue of semi-tropical swamps, on the Antarctic continent. This discovery was explained by the unproved hypothesis that the earth's axis had shifted in the past and thus the North and South Poles had moved. The other observation was the suggestion, made in 1906 by a French physicist, Bernard Bruhus, that the magnetic poles had reversed themselves in the past; but this theory was given slight attention. It was just another harmless possibility of "pure" science.

During the nineteenth century, scientists in western Europe and the United States were very creditably elucidating the history of rocks exposed on land. They staked out eras such as the Cenozoic, periods such as the Cretaceous, and epochs such as the Pleistocene. They charted the Ice Ages and established the dates of primeval events (but only in relation to one another) such as the Age of Dinosaurs and the Laramide Revolution that created the Rocky Mountains. They had enough to do without troubling about the geology of the ocean floor and moreover they had hardly any tools to do the job.

Before the Civil War, the first American oceanographer, Captain Matthew Maury, was astute enough to write, "Could the waters of the Atlantic be drawn off, so as to expose this great sea-gash, which separates continents and extends from the Arctic, it would present a scene most rugged, grand and imposing. The very ribs of the solid earth, with the foundations of the sea, would be brought to light and we should have presented the empty cradle of the ocean."

Maury, whose interest in the sea was largely limited to how it concerned the operation of ships of the U. S. Navy and merchant marine, became involved with the deep ocean through the invention of the telegraph. Samuel F. B. Morse, the American inventor who much preferred to be remembered as a portrait painter, had first demonstrated his instrument for swift electric communication in 1844. An entrepreneur named Frederick Newton Gisborne went to a New York businessman, Cyrus Field, with an idea for running a telegraph cable all the way across the Atlantic to England. Field consulted Morse and Matthew Maury, and both believed it could be done. Maury interested the U. S. Navy in the project, and a lieutenant named Berryman was sent out on a ship, the *Dolphin*, to survey the waters between Newfoundland and Ireland where North America and Europe are closest together.

Berryman's crew measured ocean depths by sounding the bottom with a piece of lead at the end of a long line—a difficult and tedious process. Berryman, whose work was done in 1853, reported that there was a "telegraphic plateau" between the two large islands. The ocean floor was not impossibly deep and the soft sediments at the bottom would enclose the cable and protect it. Maury said of this plateau that it seemed "to have been placed there especially for the purpose of holding the wires of a submarine telegraph and keeping it out of harm's way."

Laying the transatlantic telegraph cable was a saga of technology and business, it should be noted, and not of science. It was the first time men had tried to work anywhere on the deep ocean floor, and they discovered problems then that still have not been solved except in a roundabout manner. At a distance of only a few feet, the sea appears to be almost totally opaque, so that you can never see what you are doing. When the objects you are dealing with may be three or four miles down and, therefore, completely invisible, great ingenuity is required if anything is to be learned about them.

The men laying the world's first deep-sea telegraph cable planned as well as they knew how. Paying out cable from a ship, they could not tell whether it lay on the bottom and followed its contours or was strung taut over the crest of an undersea mountain in such a way that some slight strain might break it. A straight line between Newfoundland and Ireland is just about 1,000 miles, but no one could say how much cable was needed to traverse the distance along the bottom, which at one point was 3½ miles deep. No one knew what to do if the cable broke or needed repairs.

A United States ship carrying cable sailed from Newfoundland, and a British ship sailed from Valencia, Ireland, with the aim of meeting in mid-ocean to splice the two ends together. The first two attempts failed, but on August 5, 1858, the cables were united and messages began to flow from London to New York: Traffic was brisk until a failure occurred two months later. The Civil War interfered with further cable-laying, and it was not until 1866 that Cyrus Field and his partners hired the *Great Eastern,* a paddle steamer that was then the largest ship ever built. The 2,000 miles of telegraph cable with which it was loaded presented a problem, not of weight but of storage space. Part-way across, the wire broke. Rather than waste time looking for it, careful navigational fixes were taken of the position and the *Great Eastern* returned to Ireland for more cable. As it was paid out, messages were continually sent through to make certain that it was functioning. At last, the ship and the cable reached Newfoundland, where the cable was spliced with the line to New York.

Full of this success, the entrepreneurs went back to search for the lost cable. Now the *Great Eastern* had three other ships in company and carried a curious device— an object shaped more or less like a beetle with numerous large, strong hooks hanging from it. This grappling device was thirty times lowered over the side, at the well-fixed position, but without success. Then it seemed to the deck crew handling it that it had hooked onto something, and slowly they raised the catch to the surface. As the hooks broke the surface and the cable could be seen, the men on all four ships let out a great cheer. Either this disturbed the handlers or the vibration caused something to slip; at any rate, suddenly the quarry slipped loose and sank to the bottom.

Once again the cable was hooked, and at about midnight on Friday, August 17, it was hauled to the surface by the light of torches. This time the crews restrained themselves. It took the crew of the *Great Eastern* until the following evening to bring enough cable aboard to make a good splice. Now the crews let out a hearty cheer as the cable began to descend again. At last the other end was brought ashore at Heart's Content, Newfoundland, and two telegraph lines had been laid across the Atlantic. It was incredible enough that a successful cable could be laid under such circumstances, but even more remarkable that a lost object could be retrieved from the invisible bottom.

The feat was a combination of hard work and ingenuity. Today, hundreds of wonderful new tools have been invented to conquer the intransigent sea. The amazing work now going on in the ocean would be impossible without them. New tools, however, seldom work exactly as well as their designers had dreamed they would. The *Glomar Challenger*, a pioneer ship equipped with a great variety of such tools, has found that very resourceful and patient men are needed to make them perform their assigned tasks. The best machines, including the shipboard computers, are no better than the men using them.

Almost one hundred years of technological development and scientific thought had to pass, however, between the first ship to bear the name of *Challenger* and its specialized counterpart. Ocean scientists are sentimental and serious about naming their ships and the organizers of the United States Deep Sea Drilling Project (DSDP) realized they might be treading on toes when they proposed to give their new ship the same name as the world's first, and probably most famous, oceanographic vessel, H. M. S. *Challenger*. British scientists, who

had unofficial custody of the name, indicated their will-
ingness to let it be used by supplying mementos of the
first ship's globe-circling ocean expedition; these are now
proudly displayed in a glass case in the lounge of the
new ship. Among them is a manganese nodule from a
depth of 12,000 feet.

H. M. S. *Challenger* sailed out from Portsmouth, England, on her historic voyage of deep-sea reconnaissance
in December, 1872. Equipped with both steam and sail,
she was a 2,300-ton surplus vessel of the Royal Navy that
had been stripped of all but two of her guns to make
room for a seagoing laboratory. The Royal Navy had
sponsored other expeditions, such as the epic voyages of
Captain Cook, and because the climate of the times was
right, it was induced to back the *Challenger*. For a number of years there had been sporadic efforts by British
scientists to uncover the mysteries of the sea. In one of
these efforts, one of the scientists, Edward Forbes, had
dredged in the Aegean Sea, and had announced that no
life existed in the ocean below a depth of 1,800 feet. By
this time, popular interest in the ocean floor had been
stimulated by the laying of the Atlantic telegraph. A
steam engine that had been developed for hauling in the
nets of Britain's extensive fishing industry was quickly
seen to be adaptable to handling heavy equipment of
other kinds in deep water.

The *Challenger*'s commander, Captain C. Wyville
Thompson, was a science enthusiast, and his officers had
been selected for their surveying experience and interest
in scientific matters. The ship had quarters for six scientists and a staff artist. Nine rough and stormy days out
of Portsmouth, the ship made its first stop for observations in the Bay of Biscay between France and Spain.
A dredge was lowered to the bottom at 1,125 fathoms,

or 6,750 feet, and it brought up a sample of ice-cold mud, enough to chill a bottle of champagne and toast the voyage's success. On examination, it turned out to consist of globigerina ooze, a sediment composed mainly of microscopic fossil shells belonging to the order Foraminifera, that now has become of great significance to petroleum geologists.

For three and a half years, the *Challenger* sailed around the world with no other purpose than to find out what existed beneath the sea and to account for its variety. The ship traveled to Portugal and the Canary Islands, crossed the Atlantic to the sunny island of St. Thomas and north to the foggy shores of Nova Scotia, moved southward to Brazil, recrossed the Atlantic to the coast of South Africa, and headed still further south to the bleak Kerguelen Islands, and finally south again to almost within sight of the Antarctic. The explorers aboard the *Challenger* did not see land, but they did encounter icebergs and gales so terrible that they passed "many a fearful and perilous night." Heading northward once more, the ship sailed to Australia for a month of rest and entertainment, stopping off in New Zealand, the Fiji Islands, the Philippines, Japan, and Hawaii before it rounded Cape Horn and returned home to England.

No voyage of this length could take place without incident. One sailor had been put ashore in Brazil with yellow fever; another had been killed in an accident off Capetown; a third—a German biologist and the only man aboard with a Ph.D.—died of natural causes off the coast of Chile. There had been a number of desertions by members of the crew. In the course of 362 stops made to take oceanographic measurements, the ship had lost 28 thermometers and broken her dredging line no less than 11 times. Despite these mishaps, the expedition had several formidable accomplishments to its credit. One of its aims had been to find out how deep the sea really is. One of its soundings in the Pacific had reached a depth of 87,660 feet, the world's record at that time. The scientists aboard the *Challenger* had also hoped to learn what was on the bottom, whether sea water was the same in all oceans, and whether prehistoric animals still lived in the sea's great depths. They were now able to announce that there were living things in the sea at every depth; that the ocean floor was not flat, as had been supposed, but often very jumbled and rough; that all sea water was chemically the same; and that the species of marine life yet to be classified ran into the thousands. The *Challenger* report filled fifty volumes and took twenty years to publish. Some of the sediment cores retrieved during the voyage are still being studied today. In an article for *Oceans* magazine, in 1969, Dr. Peterson and Dr. Edgar summed up the *Challenger*'s achievement: "This voyage established, once and for all, the broad contours of the oceans basins, *adding the third dimension of depth. . . .* Now the *Glomar Challenger* is *introducing a fourth dimension, time.*"

Millions of miles of tracks have been made on the sea surface between the day in 1876 when the *Challenger* returned to England after her epic voyage, and our own day, when a federal register lists no less than 85 United States ships belonging either to the government or to universities devoted to oceanography. The intervening years have shown a steadily, and then rapidly, increasing interest in the sciences of the sea. A ship such as the *Glomar Challenger* could never have been built, nor the problems to which it is dedicated dreamed of, without tremendous advances in technology and knowledge since the pioneer days of the first *Challenger*.

When men began to have visions of finding out what lies *beneath* the ocean floor, they already had a very clear picture of the surface of the floor itself. Beginning with the H. M. S. *Challenger,* most research ships sent down various scoops and dredges to secure samples from the bottom. The various kinds of coring devices now in use began with a simple tool, much like an apple corer, and now include a piston core that is capable, under exceptional circumstances, of bringing up samples from nearly 100 feet beneath the ocean floor.

Aboard the original *Challenger,* all soundings were laboriously taken by dropping long, leaded lines—an occupation that frequently occupied several hours. After studying the various depths recorded by this method, Sir John Murray, principal editor of the report on the expedition, wrote in 1904 of "submarine hills," and, in 1912 he published a paper suggesting the possibility of an Atlantic mountain range.

Meanwhile, around the turn of the century, other explorers had also made hard-won soundings. Fridtjof Nansen, a Norwegian explorer who hoped to reach the North Pole by ship, deliberately allowed his vessel, the *Fram,* to be frozen into the winter ice on the theory that ocean currents might carry him to his destination. Although this did not happen, as a result of the soundings taken by the *Fram,* Nansen made the discovery of a high "sill" between the Arctic Ocean and the Atlantic that prevented the deep waters from mixing.

Prince Albert of Monaco, an amateur oceanographer of considerable accomplishments, had made numerous soundings from his yacht, the *Hirondelle,* in the Mediterranean as well as in the Atlantic. In the Atlantic, he found a deep, unexplained gash on the ocean floor, which he called the Romanche Trench.

At about the same time, the United States geologist, Alexander Agassiz, on board the *Albatross*, discovered from soundings off the Pacific coast of Mexico a remarkably shallow area that he named the Albatross Plateau. Similarly, the *Blake* of the U. S. Coast and Geodetic Survey, in taking depths off the coasts of South Carolina, Georgia, and Florida, found a broad area that was neither continental shelf nor deep sea. Scientists named this the Blake Plateau.

In 1922, the German research ship *Meteor* became the first to use a sonic depth finder that could take soundings made without lead lines while the ship was moving. Figures taken with this new kind of depth finder accumulated in sufficient quantity to show a true picture of the ocean floor. Most of the *Meteor* tracks were made in the North Atlantic and it soon became clear from its findings that a mountain range lay beneath the waves. Later soundings showed that the same range extended into the South Atlantic. This observation was confirmed by the fact that different temperature and salinity measurements were found in the eastern and western waters of the South Atlantic. Something was obviously preventing the two sections from flowing together.

With the use of this depth finder, many unsuspected ocean features soon were revealed. In the 1930's, a Danish ship surveying in the Indian Ocean found a large mountain range, which was named in honor of the Carlsberg Brewery, a sponsor of the expedition.

As depth-sounding equipment improved over the years and more and more ships employed it, the ocean floor began to emerge as a very splendid landscape, full of variety. Without ever having actually seen what they were describing, oceanographers began to speak of abyssal plains, escarpments (cliffs), seamounts and guyots (flat-topped, deeply buried mountains), fractures, canyons, and trenches.

The pace of the revolution in ocean geology begun by the first *Challenger* accelerated as new tools were developed. Probably the most important equipment to be developed had to do with earthquakes and with the discovery of the lodestone and the magnetic compass. These developments include the process known as "seismic shooting" and the instrument known as a "magnetometer." But the understanding of what deep-sea drilling means also depends on the following:

Invertebrate paleontology: The science of ancient tiny animals.

Gravimeters: Instruments that measure changes in the earth's gravity.

Heat probes: Thermometers that measure heat flow from the ocean floor.

Radiometric dating: The process of establishing the age of ancient rocks by measuring the decay of their radioactive elements.

Satellite navigation: A method of telling where on the globe a ship is drilling within an accuracy of 300 feet or less.

Sonar: Sonar beacons are dropped to the ocean floor to signal the ship's position in relation to the drill hole.

Computers: Machines that digest the sonar-beacon information and regulate the ship's propellers to keep it in position.

The oil business: Without its drilling experience, the whole enterprise would be unthinkable.

This partial list of the necessary equipment and knowledge shows some of the complexity of the whole project. For men to have dreamed of such an experiment and to have been able to physically bring it about, they have had, in the final analysis, to call upon most of the resources of our modern civilization. All of this is brought to bear in order to recover just a pinch, a speck, of buried material. For that, physically, is what they are after: the tiny specks of material from which they can tell the history of the world.

2

"The sedimentary layer on most of the ocean floor is only about 2,000 feet. We have every reason to believe that in that 2,000 feet of sediment the whole history of the earth is better preserved than it is in the continental rocks. The dream of my life is to punch that hole 2,000 feet deep and bring the contents back to the lab and study them."

—MAURICE EWING,
Director of the Lamont-Doherty
Geological Observatory

ANY HISTORIAN of "big science" is virtually certain to offend someone. Scientists are quite human and very sensitive in the matter of credits. The reward of their work is not wealth or power, but recognition, particularly by their peers, of their achievements. In something such as the DSDP in which at least 300 scientists are involved to some degree, no writer can judge which of them merits the most attention. Some scientists whose names are not on any of the official records may have contributed more than Dr. Someone who signed a voyage report. The engineers and the technical men involved also deserve great credit but their names will never appear in the scientific papers, which are written in a very formal style that rigorously excludes every hint of human beings as participants in the activity described.

No one involved with the story of the *Glomar Chal-*

lenger, however, will deny that Maurice Ewing deserves a mention in it. If nothing else, he brought the absolutely essential element of seismic shooting to the sea. Every famous man has detractors and Dr. Ewing has his, but his fame as one of the founders of American oceanography is quite secure. It was not because of sentiment that he was named one of the chief scientists of the *Glomar Challenger's* Leg I. He knew as much about the area, the Gulf of Mexico, as any scientist alive.

Dr. Ewing has never taken a course of geology in his life. He did, however, work his way through college and get his doctorate in physics at Rice Institute, in Houston, Texas, by working summers with oil company teams prospecting off the shores of Texas and Louisiana. The technique the teams used to find salt domes (small rounded peaks in the earth in which oil is likely to be trapped) was to drop explosive charges over the side of a boat and from the recorded echoes get a picture of the shape of the ocean bottom. The principle is the same that seismologists use to detect earthquakes. "If you want to sum up what I have done," Dr. Ewing says, "you can say that I have adapted the tools and tricks of the oil companies to exploration of the deep sea." While teaching at Lehigh University, Dr. Ewing wrote a paper about using the same technique to study the effects of quarry blasting, and this led to an invitation to try the same kind of surveying out in the shallow waters of the Atlantic continental shelf.

"Little was known of the geological structure of this area or of transition from continental shelf to ocean. If I could get seismic readings of it comparable to those I had been getting on land, the information might provide an approach to the fundamental question, Why are some parts of the earth continents and other parts ocean?"

With two colleagues, Dr. Ewing carried a portable seismograph (the instrument that was designed to detect earthquakes) across the coastal plain of Virginia, from Petersburg to Cape Henry, and then aboard a ship of the U. S. Coast and Geodetic Survey. Every few miles as the ship proceeded eastward it was stopped and the seismograph lowered to the sea floor. Then the three young men moved off some distance in a whaleboat to sink blasting charges to the ocean bottom and then explode them. As the charges went off, the seismograph picked up the waves of sound energy that were reflected and refracted from the earth beneath the sea bottom. The difference in the amount of time the waves took to return from the apparent bottom, the sea floor, and from the true ocean basement beneath the sediment, told the thickness of the sediment layers and gave an idea of its density. A little later Dr. Ewing was invited to repeat the same kind of experiments from the *Atlantis*, the oceanographic sailing ship that belonged to the Woods Hole Institution in Massachusetts. At that time, opinions about sediment depth ranged from as much as 12 miles to as little as 500 feet. In the area studied, Ewing discovered that the refracted waves gave a thickness that varied from 1,500 feet to 3,000 feet.

Since those early days in 1936, seismic shooting has become much more sophisticated and refined. It is no longer necessary to lay an instrument on the ocean floor or to explode the charges of TNT on the bottom. The sound waves are picked up by hydrophones attached to the ship and the sound itself generated by depth charges and, more recently, by electric sparkers and "air guns" which are a good deal safer to use than explosives. With the latest equipment, a moving ship can now receive a continuous record of the thickness of the sediment, of the buried bedrock beneath it, and of the stratification of that bedrock. This might be compared to having an X-ray motion picture of the ocean floor lying under the ship.

Eventually Dr. Ewing moved from Lehigh to the Woods Hole Institution, where he continued his seismic work and branched out into other types of oceanography. At the end of World War II, he went on to Columbia University where he was named a professor of geology, even though he had never formally studied the subject. He was made director, in 1949, of the new Lamont Geological Observatory that was being established on the 100-acre property willed to the observatory by the late financier. Under Dr. Ewing's direction, Lamont-Doherty (the last name has been added in honor of another large bequest) has become one of the nation's and the world's leading centers for deep-sea research.

Seismic shooting at sea brought a new picture of the ocean floor simply by measuring waves of sound. Waves set off by earthquakes had been studied for more than a century by seismologists in order to look deep into the interior of the earth. The sound waves employed by the oil geophysicists and then by Dr. Ewing and the oceanographers who followed are not different from true earthquakes. They are just smaller. Since Ewing took to the sea with it, seismic shooting has spread rapidly in ocean science; first at Woods Hole, then at the Scripps Institution of Oceanography in California, and today by researchers everywhere in the world. Its first great success was to show that the crust under the Pacific was just about the same as it was under the Atlantic, and that the Pacific was not, as was generally believed even after World War II, the only "real" ocean. The second major success was to show that ocean floors are different in kind from the continental crust, that the one is primarily made of basalt and the other of granite. Seismic shooting also demonstrated that the distance to the Moho (*which may be the earth's original crust*) was about 4 to 6 kilometers beneath the sea but anywhere from 25 to 40 kilometers beneath the continents. For this reason, the Mohole, if it had come to pass, would have been drilled through the sea floor rather than on land.

Seismic shooting was also useful in the investigation of the Mid-Atlantic Ridge and Rift. Seismic work demonstrated that there was very little, if any, sediment on this great mountain range with its cleft in the center and, therefore, that the feature was geologically very young. The discovery of the rift in the Atlantic—a very important event in geophysics in the 1950's and one closely associated with the Lamont Geological Observatory—was directly responsible for the thinking and planning that resulted in the DSDP and so, of course, to the cruises of the *Glomar Challenger*.

When Lamont scientists, after studying their own surveys and those of many other researchers, published their finding that a great mountain range circled the earth twice under the seas, other scientists wrote about it in amazement. Dr. Harris B. Stewart of the U. S. Coast and Geodetic Survey stated, "The discovery of this global mountain chain or, rather, the discovery that numerous previously known individual ridge systems were all part of the same worldwide system is probably the most exciting major discovery about earth science in the past twenty years."

Dr. Ewing also remarked about this invisible feature, known only through the use of instruments. "Imagine millions of square miles of a tangled jumble of massive peaks, saw-toothed ridges, earthquake-shattered cliffs, valleys, lava formations of every conceivable shape—that is the Mid-Ocean Ridge."

As if this were not enough, the Lamont people followed up by announcing that this great ridge had a rift in it, a great crack in the Mid-Ocean Ridge with very unusual, active features. The announcement of this rift started the old thoughts about continental drift rumbling once again.

The story of the discovery of the rift has been told more than once, not only because of its importance but because it is such a pleasing example of scientific serendipity, a happy accidental discovery. Bruce Heezen of Lamont, who played a very major role in the event, wrote about it in 1960, in the *Scientific American.* "In 1953 Marie Tharp of the Lamont Geological Observatory and I were making a detailed physiographic diagram of the floor of the Atlantic, based upon a large number of echo-sounding profiles. As the preliminary sketch emerged, Miss Tharp was startled to see that she had drawn a deep canyon down the center of the Mid-Atlantic Ridge."

There did not seem to be enough data at that time to pursue the matter. But when problems with the transatlantic cables began to occur once more in the deep ocean, Heezen was asked by the Bell Telephone Company to help explain the breaks that had been appearing in the system. It seemed a good idea to look at earthquakes as the cause of the trouble.

These earthquakes, of course, had been detected by seismographs. Earthquakes at sea were usually unnoticed until a network of recording stations began to be established in Europe and North America. Then it became clear that the Atlantic had frequent earthquakes. When plotted on a map, their centers turned up most often in the center of the ridge. It was in this area, too, that most of the telephone breaks had been occurring. Miss Tharp's canyon became extremely interesting.

Now enlisting Dr. Ewing in their study, the team of three looked at soundings in the South Atlantic and found that there seemed to be a rift in the Mid-Ocean Range too, and that earthquake centers could be plotted there as well. The same thing appeared to be true of the Carlsberg Range in the Indian Ocean and the Albatross Plateau off Mexico and Central America. Using Lamont's recently acquired research ship, the *Vema*, a cruise from New York, south across the Atlantic, around South Africa, north through the Indian Ocean, and into the Red Sea produced more evidence of the ridge and rift, suggesting that this was a worldwide feature.

Since the first publication of the ridge and rift idea, many surveys have been made all over the world where its appearance has been predicted. Whether they were organized to prove or disprove the existence of the ridge-rift system, the surveys seem to have shown that the feature is very definitely a fact and not fiction. Depth finders have contoured it in ever increasing detail. Earthquake centers have been found along much of its course. Seismic profiles and piston cores taken near the center have found the sediment either nonexistent or remarkably thin. Thermal probes have found high heat flows emerging from the ocean floor in the suspected areas. Readings with magnetometers have demonstrated the special kind of magnetic pattern that has come to be associated with the sides of the ridge.

As understood today, the ridge-rift system in the Atlantic looks like the long skeleton of a fish, its back broken in a number of places by great fractures that offset it, at times, from its exact position equally distant from the continents on either side. In the Atlantic, it comes ashore at a number of volcanic islands. In Iceland, a tourist can actually see the rift. It runs across the land for many miles as a wide ditch, at times as much as forty feet deep, and its expansion is actually being measured. In the Atlantic, volcanoes are associated with the rift; and on Iceland, the old volcano of Hekla has recently become active and in 1957, a new volcano, Surtsey, rose out of the sea nearby.

As the mountain range and rift proceed south through the Atlantic a number of other volcanic islands appear on or adjacent to it: the Azores, Ascension Island, St. Paul's Rocks, St. Helena, Tristan da Cunha, Gough, and Bouvet. (It should be noted that the ocean floor also has thousands of submerged, extinct volcanoes that have no immediate relation to the rift.)

At Bouvet, far south of South Africa, the mountain system turns east and then north to enter the Indian Ocean. The system in the far south here is not so well defined as it is further north, but it becomes very pronounced once more as it becomes the Carlsberg Ridge. More or less in the center of this ocean, the system splits in two. One branch runs north and west to the entrance to the Red Sea at the Gulf of Aden. Here one branch enters the African continent near the port of Djibouti in Somaliland. The great crack in the earth forms the southern end of the Abyssinian Highlands, then bends south to create the Rift Valley of Eastern Africa. Deep and great lakes such as Tanganyika and Victoria Nyanza have formed in these rift valleys and the lakes are fed by water from high volcanic peaks such as Kilimanjaro and Mount Kenya. The other branch of the rift runs into the Red Sea and splits that probably geologically very new body of water in two. In its Red Sea passage the rift is related to a unique, mysterious type of sea bottom water often called the "hot spot." At the Gulf of Aqaba, the rift goes ashore and its depression creates the bed for the Dead Sea and the river basin for the Jordan.

The other branch of the ridge from the central Indian Ocean swings wide below Australia and New Zealand, then aims toward South and Central America. Several thousand miles west of the tip of South America the range assumes the name of the East Pacific Rise and sweeps up on a broad front until it becomes the Albatross Plateau off Mexico. While passing South America two tributaries branch off toward that continent as the Chile Rise and the Nazca Ridge. Both the Chile and Nazca have been plotted with earthquake centers.

The rift begins to have a very practical effect in North America, where it has created the Gulf of California and has been spreading Baja California apart from Mexico for the last four million years. Then it enters upon the land near the mouth of the Colorado River, runs north to create the basin for the Salton Sea and Death Valley, turns left to create the valley through the San Bernardino Mountains, and then proceeds north in California as the San Andreas Fault. North of San Francisco, the rift, or San Andreas Fault, goes back into the sea at Cape Mendocino to make its last appearance on land as the Lynn Canal in Alaska. Californians who quite reasonably feel threatened by the San Andreas fault may find a peculiar pride in knowing that theirs is not just some provincial problem and that their golden state has been singled out by a force of truly cosmic proportions.

Of the various properties of the Mid-Ocean Ridge studied to prove its existence and determine its form, the study of the peculiar magnetism of its rocks was the one that led the way to even more unsettling concepts in today's revolution of geological thought. Instruments studying the magnetic variations of rocks lying on either side of the central rift have been greatly responsible for making continental drift a respectable idea.

Many children have played with magnetic horseshoes and seen how they attract steel objects such as pins. Little pocket compasses have been used by many people. The compasses, of course, point to the North Magnetic Pole. Magnetism certainly is related to electricity and it is probably produced by a core of iron-rich material spinning inside the earth. No one can really describe it or draw a picture of it, but uses were found for it hundreds of years ago. Someone, perhaps a Chinese, discovered that a sliver of a rock called a lodestone, if swinging freely on a pivot, always pointed north and down. This lodestone became the first compass by which men found their position at sea.

Navigators using their compasses in far northern latitudes found that their instruments became unreliable and no longer pointed to the geographic North Pole. Up in the high latitudes, the compass needle either swung wildly or pointed to the magnetic pole that, in recent historic times, has lain on the Prince of Wales Island, hundreds of miles from the North Pole. Outside these high latitudes, the compass dutifully pointed in a direction that could be considered true north, unless it was over a portion of the earth with an abnormally high iron content.

In 1906 Bernard Bruhus reported that the North Magnetic Pole had "failed" in times past but the matter did not seem urgent, or at least the consequences were not understood until geophysicists working in Iceland found that the lava rocks there showed that the magnetic field had frequently been reversed. Studies of the lavas showed that when samples of the same age, but from various places, were matched, their magnetic orientation was always toward the same position, but a position not occupied by the magnetic pole at the present time.

The magnetism in rocks is stable. It is locked in and cannot change unless heated to a high temperature. The locking-in of magnetism was proven by magnetometer tests on the lava flows of Hawaiian volcanoes whose eruption dates were known exactly. Then, in 1958, when it became known how to date rocks precisely by the different percentages of their potassium and argon content, Australian scientists studied their lavas to date them by this means rather than by the old comparative, geological method. The magnetic reversals in these rocks showed that all those of similar date had the same magnetic polarity. Rocks of the same age from as far away as California also had the same polarity but in neither case was it toward the present magnetic pole. Dating of ancient rocks by means of their built-in magnetic polarity shows that two hundred million years ago England was much closer to the Equator than it is today and that it has rotated considerably in relation to the modern magnetic pole. Data from India suggest that it was once far south of its present position—far enough to account for the scars of past heavy glaciation on mountainsides that can be seen today. During these last few hundred million years it can be shown that the North Pole has moved from somewhere in the mid-Pacific to its present location. A recent calendar drawn up from ancient lava flows found on land provides a timetable of 171 magnetic reversals that have been found to have occurred in the last 76 million years.

The underlying element in the magnetism of rocks is an ore called magnetite. Magnetite, in which lodestone is found, is present in many lavas and it contains a high percentage of iron. When lava flows out of its mysterious birthplace it has a temperature of 1,000°C., or more. This hot red fluid melts whatever rocks it encounters, creating a new blend of material on the earth's surface. When it cools through what is known as the Curie point, somewhere between 575°C. and 600°C., the iron in the compound becomes magnetized and lines up with the magnetic pole as it exists at that time. The polarization is permanent. Over a period of many millions of years that magnetized rock may be folded up making a high mountain, but its fossilized polarization remains the same. This polarization can be used as a magnetic tape recorder.

The instrument used to study the polarization of rocks is called a magnetometer. It can be towed behind aircraft and streamed behind ships (in which case it is sometimes referred to as a "mag fish"). Carried sufficiently far away from the vehicle so that the magnetism there cannot affect results, the magnetometer shows the polarity of the rocks over which it passes.

Applied to the rocks on either side of the Mid-Ocean Ridge, the magnetometer shows that, as one moves away from the center, the polarity of the material has had frequent reversals.

The phenomenon of magnetic reversals on the sides of the ridge has an added, curious feature. The pattern is the same on either side, a mirror image. This fact gives great support to the basic, generally-held theory about the rift, the Mid-Ocean Ridge, and the causes of their creation.

The discovery of what is happening in the center of the great mountain range of the world's oceans is probably the most important element in what Sir Edward Bullard, perhaps Britain's leading geophysicist, described as "the great increase in understanding of the present state and recent history of the ocean basins that we have gained in the past twenty years. . . . For the first time the geology of the oceans has been studied with the energy and resources commensurate with the tremendous task. It turns out that the main processes of geology can be understood only when the oceans have been studied; no amount of effort on land could have told us what we now know. The study of marine geology has unlocked the history of the oceans, and it seems likely to make intelligible the history of the continents as well."

As has been said, this great increase in understanding began with the discovery that ocean basins are different *in kind* from continents. The mountain ridge in the ocean is not made of folded, sedimentary rocks, like the Rockies or the Andes, but of basalt that has flowed up from the center of the earth.

The ridge has a crack or rift between its highest peaks and here new basalt is continually being added to the ocean floor. The earthquakes, the heat flow, the lack of sediment, the frequent volcanoes all demonstrate this. The sides of the rift are continually moving apart from one another, for reasons that are still hotly disputed, and the rift would widen if it were not being refilled. As the rocks on either side move away, they cool and assume the orientation to the magnetic pole that exists at the moment they pass the Curie point. The pole has reversed itself many times over the ages but the rocks on the sides of the ridge remain the same and they can be identified by use of a magnetometer. Dr. Robert Dietz gave this phenomenon the name of "sea-floor spreading," and *Glomar Challenger* reported on its voyage to Dakar, its the process has been called this ever since. When the most noteworthy finding was that it had demonstrated the fact of sea-floor spreading.

If the sea floors are spreading, of course, they must be spreading to some place, an observation that leads immediately to the idea of continental drift. The sea floors must be spreading the continents apart.

The new geology, however, is not the only reason for today's revival of the idea first formulated by Sir Francis Bacon in 1620. Strong support for the theory of continental drift has come from a number of sources in recent years.

There is, for one thing, the fact that no truly ancient ocean sediments have been found and that no scientist expects that any will ever be found to be more than 200 million years old. To explain the disappearance of everything older than this, one may assume that the drifting continents have acted like bulldozers and buried all evidence of any former seas.

The study of land geology has produced a story that shows that ancient climates of the present continents were far different from the way they are in the present. This branch of science, called paleoclimatology, assumes that the earth's axis of rotation has not changed appreciably and that the modern Equator has remained in the same position. The belts of winds, of dry and wet climate, would thus have always been as they are now. If this assumption is true, then today's rocks should show the effects of ancient weather. Thus the geology of Europe shows that in Permian times, about 230 million years ago, the land lay in the tropical zone and northern Europe was covered by a desert. In the Pennsylvanian age of North America, 250 million years ago, the equator ran diagonally across the continent from Newfoundland to the Gulf of California. The major North American coal beds were formed at this time. Later, in the Triassic period, 205 million years ago, North America was far enough south to give Canada a tropical climate and allow coral reefs to grow on the edge of the Arctic Ocean.

From other evidence of land rocks, red sandstones of comparable age in Great Britain and Arizona show by their magnetic polarities that North America and Europe in the extremely distant past were much farther away than they are today. Continental drift may mean that the land masses have moved back and forth more than once and have always been in a state of motion.

Many comparisons of rocks from Brazil and West Africa have shown great similarities in the formations found in both places. This is also true of a number of animal fossils, relics of ancient animals that were identical— creatures who could not have swum across the present salt seas of the South Atlantic.

The recent discovery of tropical, freshwater, amphibian and reptile fossils in the mountains of Antarctica has been further very strong proof of the drift idea. Scott and Shackleton had discovered coal beds in the Antarctic and geologists had found the imprint there of *Glossopteris*, a large plant that also thrived millions of years ago in South America and South Africa—all showing that the frozen continent must once have been warm. These remains of the past were from plants, however, and the seeds might have floated over the ocean and taken root at a time when the Antarctic, although still in its present position, might have been warm because the earth's axis had shifted and the South Pole lay elsewhere. The fossil bones of animals whose relatives had also existed in South Africa and South America ruled out this kind of explanation. No matter how the South Pole might have shifted, the animals would never have crossed the broad, salty seas. They must have reached Antarctica when the land touched one or the other southern continent.

Another confirmation of drift came recently when a Columbia University scientist, whose findings were confirmed by scientists from 11 other nations, showed that the South Pole of 450 million years ago has now become the Sahara Desert. Great parallel grooves in the Sahara rocks showed that the land had once been under a polar ice cap, a southern ice cap since the grooves moved from south to north. No ice cap could ever have come as close to the Equator as this Sahara region is today. The striated rocks were dated radioactively and even more firmly by the discovery of fossils of trilobites which became extinct just after this time.

Computers have also been used to add weight to the drift hypothesis. Sir Edward Bullard has fitted the present edges of the continental shelves of Europe, Africa, North, and South America together by the use of immense numbers of statistics. His resulting picture of a time long ago shows how all the continents fitted together almost perfectly, with only a few overlaps and a thin patch of open water between Iceland and Spain, which then runs down between Nova Scotia and Morocco. Sir Edward's reconstruction also leaves much open water where the Gulf of Mexico and Mediterranean Seas lie today.

A similar reconstruction has been made of the fit between the Antarctic and Australia by Robert Dietz, now of the Environmental Science Services Administration, and Walter Sproll. Their data seems to show that these two continents separated about 40 million years ago. Before the split began, the southeastern end of Australia, including Tasmania, fitted snugly into the Ross Sea, and the concave Great Bight of Australia easily joined the convex outline of Wilkes Land in Antarctica.

These essays at putting the continental jigsaw together again are not the only ones that have been tried, nor will they be the last. Each of these attempts, however, gets a little closer to the mark. The long process of finding out how the continents related to each other at various times past has just begun. A new theory, called plate tectonics, that was just being talked about as the *Glomar Challenger* began its first cruise, may help simplify the many problems that arise when one accepts the fact that, almost certainly, our world today just is not what it used to be.

Digging long holes of small diameter into the debris that lies beneath the sea may not seem glamorous when compared with the heady game of juggling whole continents and millions of years, but the two activities are quite immediately connected.

But where did the drilling idea come from in the first place? As far as the written record goes, it seems to have started with the most revolutionary scientist the world has ever known. He was a very modest man, an invalid called Charles Darwin, and he mailed a letter about drilling to an American friend in 1881.

3

"I wish that some doubly-rich millionaire would take it into his head to have borings made in some of the Pacific and Indian Ocean atolls and bring home cores for examination from a depth of 500 to 600 feet."
—Letter from CHARLES DARWIN to ALEXANDER AGASSIZ

THE MAN whose name is synonymous with evolution belongs to the story of deep-sea drilling for several reasons. Darwin's theory on how species originated is the backbone of paleontology, the science of fossils from the ancient past. And the paleontology of the microscopically small invertebrates is a major method of dating ocean sediments. Many thousands of invertebrate species, a good number of them extinct, have well-known places in the geologic time scale, and the evolutionary form in which they appear in a particular core gives a fairly exact indication of when the sediment it contains was deposited.

While sailing aboard the *Beagle* and studying the living species that led to his great theory, Darwin also observed the atolls and the coral islands his ship passed in its travels. Atolls are unique in that they are low, circular reefs whose rock-like structures are dead coral, animal shells that have cemented together. Darwin had an idea that these atolls were formed around volcanoes. The living corals attached themselves to the sides of the volcanoes, at sea level or not far below. The sea might rise and the volcano sink, but the living skeletons kept building on the skeletons of the past, maintaining the island long after the original volcano had disappeared far beneath the surface. The depth to which Darwin thought the millionaire's drillers would have to go was woefully shallow of the mark. Trying to find the old volcano under the atoll, which Darwin believed had to be there, the Royal Society of Great Britain funded a hole in the Ellice Islands of the Pacific in 1897 and the drillers found nothing more than coral even at a depth of 1,140 feet. Japanese scientists, in the 1930's, tried out the same idea on an atoll and went down to 1,416 feet without proving Darwin's idea. Just before the A-bomb tests on Bikini, in 1947, three Americans—Harry Ladd, Gordon Lill, and Joshua Tracey—were involved in a drilling project. They reached to a depth of 2,556 feet but still did not encounter volcanic basalt. Finally, in 1952, soon after the first H-bomb explosion on Eniwetok, Harry Ladd led a drilling project that finally found basalt at a depth of 4,222 feet. The old volcano on which Eniwetok atoll was formed had subsided this much since coral first anchored themselves to the sides of its cone. Drilling finally proved the truth of Darwin's theory about how atolls were formed.

Following the discovery that enormous fortunes could be made from the quantities of oil beneath the earth's surface, practical men began to develop better and better techniques for drilling and recovering it. In their search for oil, prospectors learned to call upon scientists for help —geologists to tell them where oil was most likely to be trapped and paleontologists to study the cores brought up by drilling to see if oil bearing formations had been reached. This early partnership between science and the oil business has helped that industry become one of the most progressive in the country in the area of scientific research. Recognition of this relationship led a visionary geologist, Dr. T. A. Jaggar (who, among other things, was one of the founders of the study of volcanology in the Hawaiian Islands), to propose, in 1939, that the oil industry become involved in ocean drilling at Honolulu. A newspaper interview from the *Honolulu Advertiser* reports him saying: "Let's take two or three of those old warships which are rusting in idleness at the wharves. Let's put the best brains of the oil industry to work on the development of a new kind of drill for boring beneath the ocean's floor. Send them to deep water and tie the boats together. Then we can put down drills to get samples of the rock beneath the mud. That's what we want. We'll have a start, then, for solving what is still an unanswered geophysical problem.

"We spent millions for an observatory to study the craters on the sun and moon. That's very commendable. But we have studied only a small number of the craters on earth. We know very little about the world we live on. We must strike the imagination of some capitalist who could finance an expedition of this type. Perhaps a syndicate will be convinced yet that oil deposits lie beneath the sea; maybe greater fields than those they have found on the tidelands."

A few years later Dr. Jaggar followed this up with a manifesto titled "Core Drilling Under the Oceans." In it he said that "The oil industry alone is intelligent enough and wealthy enough to carry and man the whole proposal, doing the diplomacy, laying out the blueprints, finding the engineers, and prospecting the world oceans for beginning shallow and ending deep. The oil geologists are enterprising enough to swing their companies and see the vision. . . . It must not be permitted to fail, for that would leave geology a speculative science as before, surviving by continental anatomy, when its real function is global exploration of the two-thirds of the surface that is under the sea." Dr. Jaggar's vision is now being realized, and the oil companies are being very cooperative in bringing the goal about, but the financier is a new and unexpected man, Uncle Sam, who only began to underwrite science in a major way after the experiences of World War II. It is a horrible but well-recognized fact that today only war, or the threat of it, has produced a Midas source of funds for science and technology.

When the U. S. government became known as the grand provider of money for science, it immediately began to receive many proposals for research. Some of the best of these ideas went to the Office of Naval Research, for the Navy was very successful in getting the Congress to vote it funds to support basic work on the frontiers of knowledge. Scientists in the Office of Naval Research, hired to evaluate the proposals received, found it difficult to classify some of the best of them. Many of those reaching the Geophysics Branch did not fit any pigeonhole. The origins of revolutions are said to be hard to trace, but early symptoms of the revolution in geophysical thought clearly appeared at this lower level of the bureaucracy during the years just following the war. So much of what was sent to the Geophysics Branch of the Office of Naval Research belonged to none of the old-fashioned categories of thought that it could only be classified as miscellaneous. This was not a very tidy method, but one that indicated that the scientists, known for asking the most profound questions about the whole world itself, were basically seeking new rather than routine means of solving the problems they proposed.

The scientists involved in processing the ideas were, of course, themselves caught up in the ferment of that time. It was an era of discovery and a time of a new hypothesis, which is continuing to this day and on the brink of producing a grand synthesis in which all the forces shaping our planet are united in one simple, elegant explanation. The Navy geophysicists probably did not, at the time, see with crystal clarity what was causing their classification predicament, but they did think their problem was rather funny. Two of them, Gordon Lill and Carl Alexis, are now credited with the idea of giving all the unclassifiable proposals to a group that they then created called the American Miscellaneous Society. As is usual in government, this society quickly became AMSOC.

AMSOC was a very unofficial club that soon attracted a number of America's brightest scientists. It had no meeting place, no dues, no membership list, no publications. It did, however, have special groups such as the Committee for Cooperation with Visitors from Outer Space and the Society for Informing Animals of Their Taxonomic Positions. The meetings of AMSOC, whenever they might happen, gave scientists in quite different fields a very relaxed atmosphere in which to exchange information and ideas. In an AMSOC group, a man might suggest a thought, however wild, that he would hesitate to offer in a presentation before a formal body of scientists. AMSOC was exclusive, as a successful club must be, but at first not aware that it was exclusive. Its fun was doomed by the quality of its membership. They were too intelligent and too well-trained as scientists to be able to play for very long. One of the ideas rather lightly proposed grew to very serious proportions: deep drilling through the earth's crust down to the mantle.

Meanwhile, independent of AMSOC, the idea of drilling holes in the earth's surface continued to make appearances. In 1953, Dr. Maurice Ewing made a tour around the country as guest lecturer for several distinguished societies of geologists. His main topic was the desirability of drilling into the sediments and his main goal was obtaining support for the project. He thought that the Lamont Observatory's new ship, the *Vema*, might be able to handle the drilling and that the initial cost might be one million dollars. In a 1954 New York *Herald Tribune* interview he said he felt that even 500 thousand dollars could do the job and "punch that hole" he had been dreaming of.

In 1956, Dr. Frank Estabrook, a scientist with the U. S. Army, published an article in *Science* called "A Geophysical Research Shaft." Dr. Estabrook offered a number of arguments for drilling such a shaft; among other matters, it might explain the origin of the earth's magnetic field, show whether the earth is getting warmer or colder, explain the continents and ocean basins, and tell what the earth's mantle was composed of. In addition, many unexpected discoveries were very likely to be made. No public exception was taken to this proposal but no one rose to promote it.

The subject of deep drilling turned up again, in March, 1957, after a meeting of the National Science Foundation —the government agency created to distribute funds for basic research to private institutions and individuals. The NSF's committee on earth sciences had been considering proposals, 65 of them, all well thought out and full of merit, but none of them were the least bit exciting. Later, two of the committee members, Harry Hess of Princeton, a geologist, and Walter Munk, a geophysicist, complained that their fields would make no advances with projects so lacking in audacity. Hess and Munk tried to think of the biggest project imaginable to move the earth sciences. Why not dig a hole down through the earth crust down to the mantle?

The men talked it over: the exciting possibilities, the very large problems. Perhaps the big oil companies would be interested. Hess and Munk parted after agreeing to bring it up at the next meeting of their AMSOC friends.

The following month, an informal AMSOC group met for breakfast at the home of Walter Munk in La Jolla. The host was on the staff of the Scripps Institution of Oceanography there. Munk and Harry Hess led the talk around to their idea of deep drilling to the mantle. The AMSOC members present, sitting on a sunny patio lookout toward the blue Pacific, were enthusiastic. None of those present knew exactly the closest distance to the Moho (or mantle) and none of them knew just how far the oil drillers had ever dug. They were not aware of Dr. Jaggar's Hawaii proposal, Dr. Ewing's lecture tour preaching the drilling gospel, or of Dr. Estabrook's magazine article proposing geophysical drilling. They did know about the drilling that had been done on atolls and they did consider making the hole from a floating platform, but no one raised the question of how the project might be financed. They named Gordon Lill, who had founded AMSOC, as chairman of the brand new drilling committee created only moments before; Walter Munk and Harry Hess as members, since it was their idea; Joshua Tracey and Harry Ladd, because they had drilled on the Marshall Islands; and Dr. Roger Revelle, who was the director of Scripps, because he was a man of vision and had worked to get the first holes drilled at Bikini. Then they adjourned and went down to Dr. Revelle's house to tell him all about their idea. Later that month, the new committee had their first meeting at the Cosmos Club in Washington and asked Dr. Ewing, who happened to be passing through, to join in their talk. He did so and became another committee member before he left. Dr. William Rubey of the Geological Survey and Dr. Arthur Maxwell, then head of oceanography for the Office of Naval Research, joined the committee at about the same time.

When this distinguished group asked the NSF to give them working capital to make a study, the organization replied that it could not deal with any group as informal as a Miscellaneous Society. Could such a body of well-known scientists get the backing of a properly recognized organization?

Meanwhile, at an international meeting of geophysicists in Toronto, under the initiative of AMSOC members, a resolution was introduced inviting all nations to consider the idea of drilling to the Moho (or earth's mantle). A Russian delegate rose to announce that "We already have the equipment to drill such a hole. Now we are looking for the place." This Russian sally was to sting the pride of the American oil industry and make them wonder just how well *they* could do such a job. An American oil driller, John Mecom, who had the current world's record for a hole 22,570 feet deep in Louisiana, was present at the next meeting of AMSOC several months later.

Then Harry Hess of Princeton, who was a member of the National Academy of Sciences (NAS), went before that organization with the request that they sponsor AMSOC and receive NSF funds so the project could be studied. Dr. I. I. Rabi, the Nobel Prize winning physicist who was a member of the Academy's governing board, expressed a sentiment with which other members seemed to agree. "Thank God we're finally talking about something besides space." The AMSOC committee came officially under the wing of the NAS, a move that gave them much needed respectability.

But then, with the opportunity offered by the presence of two hundred scientists at a meeting of the American Geophysical Union in Washington, the NAS announced a special meeting at which the idea of drilling to the Moho would be discussed. Gordon Lill began a speech pointing out the benefits that would be derived from the project but heard loud objections before he could finish.

The first complaint was that the mantle might not be the same everywhere in the world and so one hole would prove nothing. To this Harry Hess replied that while not as much could be learned from one hole as from two or a hundred holes, it was necessary to begin someplace.

Another objection was financial. If millions of dollars were set aside for this drilling project, it would rob other valuable research projects of funds for years to come. If such an amount of money were divided among many institutions, they would all be able to do better work. Roger Revelle answered this by saying that he supposed Ferdinand and Isabella heard much the same thing when they were considering Columbus' idea; someone surely had said that a crazy idea such as sailing west would prove nothing, or that the whole Spanish fleet would be improved if the money proposed for Columbus would be put into better sails and better rigging so that the whole Spanish Navy could sail a half a knot faster.

The third argument against the drilling was that it was impossible anyhow at the present time. Why not forget about drilling in the sea, which no one knows anything about, and wait a few years until drilling techniques on land had developed further, as they certainly would.

With this, an engineer from the Union Oil Company was given the floor and he began to show a movie. His film was of the *CUSS I*, a craft owned by a combine of the Continental, Union, Shell, and Superior Oil companies.

The *CUSS I* was shown off the coast of California, with a complete oil drilling rig aboard, making a hole in the ocean floor through 200 feet of water. Until this moment, the oil companies involved had kept it almost a complete secret from their commercial competitors, but now they were willing to show that a floating drilling platform already existed and that it worked.

The movie broke down all opposition. If American technology could already go this far, what was to stop it from going all the way? The enthusiastic scientists voted unanimously in favor of a resolution approving the drilling idea.

Then the project began to move forward very rapidly, particularly for something so bold, so revolutionary, so untried. The NSF put up funds, existing drilling ships around the world were studied, possible drilling sites were examined, and the news spread in waves through the scientific community and through various industries that might benefit. An article about the matter, in terms both of the science and technology involved, appeared in the *Scientific American* in April, 1959. Titled "The Mohole," it was the first story to use that word.

What first excited men about the Mohole idea was the prospect of getting to the earth's mantle, that mass of dense material that seems to exist as a layer covering our planet beneath its crust. This mantle was first demonstrated scientifically through the study of earthquake waves. A Yugoslav geophysicist named Andrija Mohorovicic had shown that shock waves from such phenomenon as earthquakes suddenly accelerate in speed as they pass from the crust into the layer underneath, the mantle. The theory has been demonstrated many times since the first discovery, and now the boundary between the crust and upper mantle is called the Moho in honor of its discoverer, Mohorovicic.

As has been said, no one knows what the mantle is made of, even though lavas that presumably originate underneath it often come to the surface. Lavas differ so much from one another that they must become contaminated as they pass through the crust.

As people will when arguing a point, enthusiasts for the Mohole found many more reasons for getting to the Moho besides obtaining a sample of mantle material. In the grandest view, one or more Moholes might tell scientists and the rest of the world the answers to the questions: What is the age of the earth? What is the age of the ocean and the earliest sediments? How old is life itself? Why do the oldest ocean sediments seem so geologically young? Does oil exist beneath the deep sea? Can the motions of continents, if they have ever moved, be tracked in the sediments? What is the true structure of the ocean crust? Is it more complicated than shown by seismic shooting? Is the Moho really an abrupt change or a gradual one? Is the Moho the original surface of the earth? Where does the earth's heat come from? Are the deep rocks more or less radioactive than continental rocks? What will deep rocks show about magnetic reversals?

These questions can be expanded on but, in the light of what followed, it may be noted that many of them relate to the ocean sediment and not the deep mantle itself. The observations of sediment were not the original motive for the drilling nor could they possibly all be made through a single Mohole.

The enthusiasts recognized this and knew that no one site on earth would satisfy all the requirements. The first site would have to be a compromise and, finally, several sites would have to be drilled before the Mohole could fulfill their expectations. Site selection became a major enterprise. Beyond satisfying all the scientific criteria, the best site would have to have minimum interference from ocean currents, the finest weather all around the calendar, and be reasonably near a good operating base because the first Mohole might take a year or more to drill.

As matters stood at this stage, outside of the government, two major United States oceanographic institutions were the most actively involved in the Mohole project, one based on the Atlantic Ocean and one on the Pacific. Dr. Maurice Ewing directed the Lamont Observatory in New York, while many of the other interested scientists were connected with Scripps in California. Both organizations were willing to send out ships to reconnoiter the best site. Willard Bascom, a marine engineer who once worked at Scripps, had been engaged in the Pacific during the International Geophysical Year, but he returned to Washington just as AMSOC was being accepted by the National Academy of Science and, as he says, "became an enthusiastic advocate of oceanic drilling. I vowed to become associated with the project somehow and before long was invited to become its part-time executive secretary."

Bascom proposed that a test drill be done first to get some experience with the problems involved in the big idea. The *CUSS I*, which had been so instrumental in convincing the geophysicists that the project might work, was based on Los Angeles and it seemed a sensible idea to begin with it as a test of capability. Thus the initial test was done in 300 feet of water eighty miles south of

Los Angeles harbor. For this, the *CUSS I* was outfitted with enormous outboard motors. Deep-moored buoys were anchored around the chosen site and, by sonar, the ship centered itself in position between them using the outboard motor screws for maneuvering. Five holes were dug in an actual water depth of 311 feet, and one of the holes reached 1,035 feet into the sediment.

With the logistic needs of *CUSS I* still in mind (the platform, after all, had only been designed for drilling on the continental shelf fairly close to port), the men then decided on a more challenging test site that still would be within 300 miles of Los Angeles. Dr. Russell Raitt and Dr. H. W. Menard of Scripps had both surveyed the waters around Guadalupe Island, off the West Coast of Mexico, in Scripps ships and they strongly recommended the area for the work. It had about the most ideal weather in the Pacific, the currents were negligible, and the ocean crust was suitably thin. Led by Bascom, a party aboard the Scripps vessel *Orca* studied the area in more detail and felt that it was the best possible choice for the next step.

At the same time as Raitt and Menard were surveying around Guadalupe Island, another small task force was engaged in similar work north of Puerto Rico. This area had already undergone extensive oceanographic study, but this time it was being looked at in a multiship re-fraction survey that would give a picture of the crust in that area in three dimensions to show its suitability for a test Mohole. With the sponsorship of the NSF and the Office of Naval Research, Dr. Ewing sent the Lamont ship, *Vema*, along with the *Bear* of Woods Hole, *Hidalgo* of Texas A & M, and *Gibbs* of Columbia University's Hudson Laboratories to do a month of oceanographic measuring. The water north of Puerto Rico is 18,000 feet deep and the depth to the Mohole is relatively short, but the deep ocean currents are not particularly well known and the area is notoriously subject to hurricanes. The region did not win out as the location for the first site.

In April, 1961, the *CUSS I* was taken to a position 40 miles east of Guadalupe Island, where the water measured 11,672 feet—suitably deep for the first true ocean drilling experiment. The *CUSS I* was held in position by the four outboard motors, with sensing done by sonar and radar on taut-wire buoys that were sunk to the bottom. Work began as if geologists had said the ocean floor was bursting with crude oil. The author, John Steinbeck, was a supercargo aboard, writing a report for a May, 1961, issue of *Life* on the operation.

"The big drill bit with $8,000 worth of commercial diamonds first, then bumper subs, and then the drill string started down. Wonderful to watch these drillers work. They are big powerful men. They carry their arms forward and their arms swing in unison. This is always true of men who lift and push at heavy things.

"The deck is heaving and pitching. The men step like cats. There is nothing clumsy about them, and as the steel sections of pipe rise and are screwed together and lowered, the drillers move with the timing and precision of a corps de ballet. They would throw me overboard if they knew I said this or thought it. But it has to be. If one man makes a clumsy or ill-timed move, someone may be killed by the swinging steel. They depend on one another for their lives. Also they have enormous pride in their work."

These men that Steinbeck described are called roughnecks and they learned their skills in the oil fields of the American Southwest. Taking their tough art out to the deep sea, just for the service of science, was quite a

remarkable experiment in itself. It takes a bright man to
work on a drilling crew, however, and even without
degrees in geophysics, the men working on oceanographic
drilling very soon become interested in the objective
behind all their activity.

The Guadalupe Island test was like all others in that it
had its share of troubles. The *CUSS I*, however, did successfully hold position for three weeks during which
time it dug five holes, one of them 601 feet into the sediment. Basalt was penetrated for the first time, far out at
sea, 44 feet of it. The gray-green ooze of the sediment
turned out to be middle Miocene, somewhere around
30 million years old. No bottom samples had ever before
been taken in such deep water. There had been worried
questions about the drill pipe. No one had ever had any
reason before to suspend more than 12,000 feet of it
below a ship, unsupported on the sides by a hole as it
would be on land. But the drill string held. As the *CUSS I*
was towed back to Los Angeles, President Kennedy sent
a congratulatory message and called the test "a historic
landmark." So drilling in the deep sea was not just the
nutty idea of some wild-eyed scientists.

With the tests at Guadalupe behind them, the Mohole
men began to think about the design of a vessel to be
built for the great project itself, utilizing what they had
now learned. They considered "the effect on morale of
food, privacy, noise, and a posted plan of operation,"
suggesting that life aboard the *CUSS I* had been a bit
rigorous and uncertain. They planned "more mechaniza-
tion and delicate control" than existed in conventional
oil drilling. Dynamic positioning needed to be improved.
The "men rapidly accustomed themselves to working on
a heaving platform and after a few days ignored the
surrounding ocean and saw only the microcosm city of

the ship." The new vessel "should be able to operate safely in the rigorous environment of the ocean." The planners "believed no danger of a blowout existed as it did at an oil well because of the great depth and volume of water. The problems of building a ship for drilling to 35,000 feet are monumental." They suggested first an experimental, modest ship to reach downward of 20,000 feet with a drill bit.

It would be several years before any more ocean drilling could be undertaken by Project Mohole.

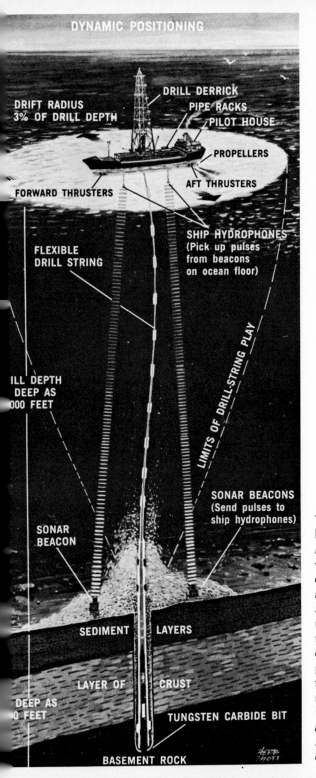

DYNAMIC POSITIONING

DRILL DERRICK

PIPE RACKS

PILOT HOUSE

DRIFT RADIUS 3% OF DRILL DEPTH

PROPELLERS

AFT THRUSTERS

FORWARD THRUSTERS

SHIP HYDROPHONES
(Pick up pulses from beacons on ocean floor)

FLEXIBLE DRILL STRING

LIMITS OF DRILL-STRING PLAY

ILL DEPTH DEEP AS 000 FEET

SONAR BEACONS
(Send pulses to ship hydrophones)

SONAR BEACON

SEDIMENT LAYERS

LAYER OF CRUST

DEEP AS 0 FEET

TUNGSTEN CARBIDE BIT

BASEMENT ROCK

The drawing at left shows how the *Glomar Challenger* remains on station while drilling in water depths up to 20,000 feet or almost four miles. Keeping an exact position involves using a system of pulses from sonar beacons on the ocean floor which are picked up by the ship and fed into a computer, automatically adjusting the thrusters to keep the *Challenger* precisely on station. *Scripps Institute of Oceanography*

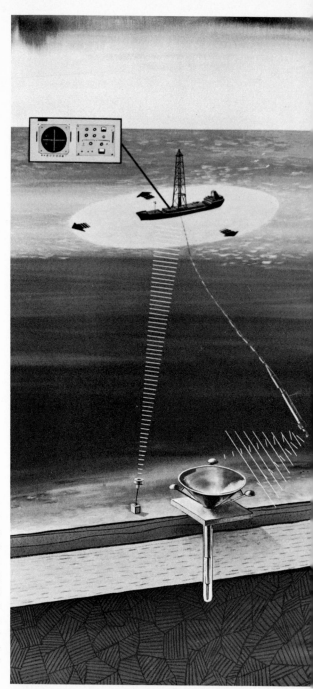

Project engineers responded to the need for reentry capability by designing the above system. When the drill bit becomes dull and useless, instead of having to abandon the drilling site, new bits can now be guided into the core hole using a scanning sonar probe and the reentry cone. *Scripps Institute of Oceanography*

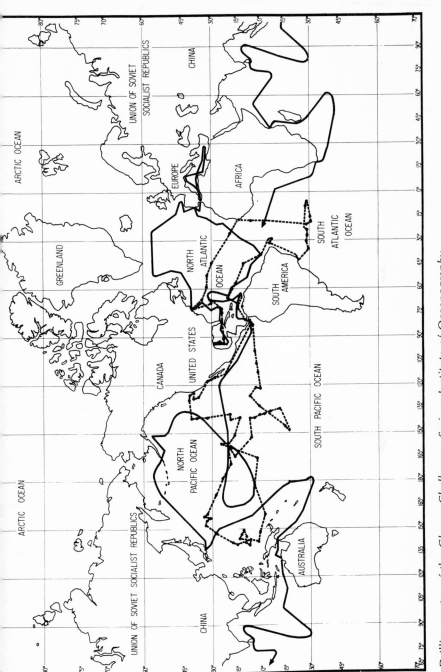

Drilling routes of the *Glomar Challenger*. Scripps Institute of Oceanography

Pacific Ocean floor. © 1969 *National Geographic Society*

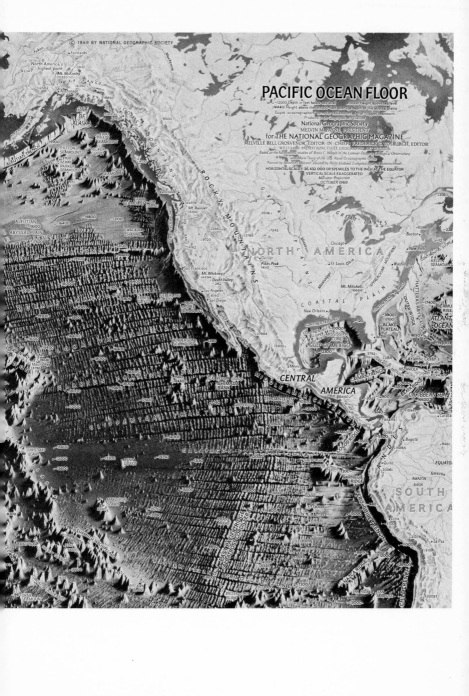

Atlantic Ocean floor. © 1968 National Geographic Society

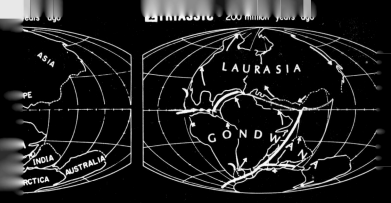

... years ago

ASIA

PE

INDIA   AUSTRALIA

RCTICA

2 TRIASSIC – 200 million years ago

LAURASIA

GONDWANA

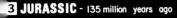

3 JURASSIC – 135 million years ago

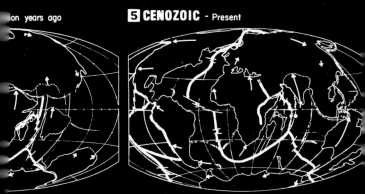

...ion years ago

5 CENOZOIC – Present

...ng how the earth's crust has broken up into vast ''plates'' and
...d masses and oceans have drifted in the last 200 million years.
...ork Times Company. Reprinted by permission.

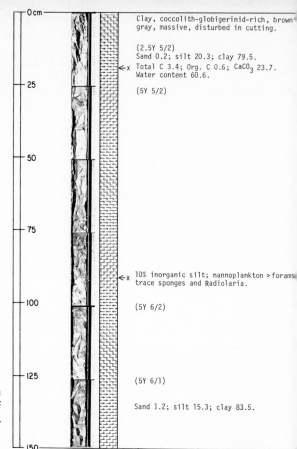

Clay, coccolith-globigerinid-rich, brown gray, massive, disturbed in cutting.

(2.5Y 5/2)
Sand 0.2; silt 20.3; clay 79.5.
←x Total C 3.4; Org. C 0.6; CaCO₃ 23.7.
Water content 60.6.

(5Y 5/2)

←x 10% inorganic silt; nannoplankton > forams trace sponges and Radiolaria.

(5Y 6/2)

(5Y 6/1)

Sand 1.2; silt 15.3; clay 83.5.

A typical core drilled from below the floor of the Gulf of Mexico. *National Science Foundation*

This scanning electron microscopy picture, magnified one hundred times, shows the skeletons of Radiolaria in cores taken in the Pacific. While alive, these tiny organisms live near the ocean surface but when they die, their skeletons fall to the sea bed and become part of the sediment. With careful study, scientists can discover how long ago they lived and from that, learn much about the history of the oceans. *Scripps Institute of Oceanography*

4

"The Deep Sea Drilling Project and the former Mohole Project are unrelated except for certain similarities in technology and their individual relations to the upper mantle program."

—Release to the press

AFTER the Guadalupe Island test, the sea floor was never disturbed again in the name of Mohole. The anatomy of why the project failed lies somewhere in the human equation because technologically it was quite possible. The Mohole idea certainly caught the popular imagination, perhaps because of its apparent simplicity. Even today, a number of years after Congress gave it an official stiletto, when some people hear of the DSDP and the *Glomar Challenger*, there seems to be a conditioned response and they ask "Mohole?"

Most of the men involved with it are now deeply engaged in the new operation, but the Mohole itself sank somewhere between Massachusetts, New York, Washington, Florida, and the West Coast. In its failure there ought to be a lesson for everyone who would rely on the federal government for support.

With hindsight it can now easily be said that a dilemma in objectives existed right from the start. Men like Dr. Ewing were fascinated with the story waiting to be unfolded in the sediments, but once the basalt at the basement had been reached they could see no real reason to continue drilling. If they went along with the idea of drilling down to the mantle, it was as a means to an end. Even as the tests were being carried out at Guadalupe, some scientists, enlivened by the idea of deep drilling, went ahead on other programs more nearly tailored to their own particular concerns.

As the prospects for significant achievement brightened, something began that a neutral observer might call a "struggle for power." The geologists whose main interests lay in the sediments were impatient that the construction of a Mohole platform would take so long and they were well aware that mantle research, when the platform was completed, would have a priority over the work they wanted to perform. Sediment drilling seemed to be possible even then, without the development of entirely new equipment, but this would be too complicated and costly for one institution to carry out on its own.

Not many months after the Guadalupe tests, Dr. Cesare Emiliani of the Institute of Marine Science, University of Miami, suggested that an existing ship be chartered for drilling in the Caribbean and nearby Atlantic. This idea stimulated the formation of a committee called LOCO, which included two scientists each from Miami, Lamont, Woods Hole, Scripps, and Princeton. Princeton had no oceanographic, sea-going capabilities but it did have Dr. Harry Hess, the geologist who had pioneered so much deep-sea work, such as discovering the flat-topped seamounts in the Pacific, which he named "guyots." LOCO never got far off paper, foundering on the allocation of responsibilities between the interested institutions. Then Dr. Ewing, J. B. Hersey, and Dr. Roger Revelle tried again with a corporation named CORE that would go ahead with the sediment part of the Mohole program. This group did not last long because it lacked any outside support.

Then the University of Miami received a grant from the National Science Foundation that allowed them to use a research vessel named the *Submarex* for drilling on the Nicaragua Rise, off Jamaica. This shallow water coring and drilling work was mildly successful in spite of bad weather. A story exists that during this period Dr. Ewing, restless at the lack of progress, had been talking with the oil companies about raising $15 million and having Lamont go ahead on its own. Whether this is true or not, the scheme did not materialize.

At last, in the beginning of 1964, while Mohole still held the public attention, the four most interested research institutions, Miami, Lamont, Woods Hole, and Scripps, got together for one more try at building a substantial organization. In May of 1964, they signed a formal agreement called JOIDES (Joint Oceanographic Institutions Deep Earth Sampling) to cooperate in deep drilling. The attempted alliance of independent powers, as happened so often before in history between nations that would like to cooperate, had always failed because the institutions involved did not want to give up any of their own rights. With JOIDES, however, it was agreed that only one institution would be in charge of any particular project and this institution would be responsible to whomever supplied the funds (in this case, the United States government). Based on such an understanding, the coalition had a chance of succeeding and, in fact, is still thriving today.

Early in 1965, it was learned that the *Caldrill,* a drilling ship with dynamic positioning, under contract to the Pan American Petroleum Company, was about to sail from its home base in California to the Grand Banks of Newfoundland for oil exploration in the summer of that year. JOIDES offered to rent it during its passage. Pan American generously agreed to pay the moving costs of the ship from the Pacific to its scheduled field of operations in the Atlantic, even though JOIDES proposed using it somewhere along its track to drill holes for scientific purposes. JOIDES secured funds from the NSF to pay for one month's drilling, and the Lamont Observatory was designated to administer the program. The JOIDES group selected the Blake Plateau for the *Caldrill*'s pause to drill research holes.

Under the immediate direction of R. D. Gerard of Lamont and John Schlee of Woods Hole, the *Caldrill* made six penetrations of the Blake Plateau, beginning in fairly shallow water 22 miles off the coast at Jacksonville, Florida. From there it moved east in steps until it was 250 miles offshore. Here the depth was about 3,500 feet. The deepest hole drilled into the sediment measured 1,050 feet and this set a deep-water record that lasted for three years. Living conditions aboard the *Caldrill* were primitive but its performance was most gratifying to the scientists watching. They appreciated not only the technical feat but the fact that the four institutions had managed to do something successfully together.

The *Caldrill* operation did not make banner headlines and it did not even solve the scientific problem of the Blake Plateau. In the political world, however, it was a quietly smashing success. The men in Washington who control the funds were impressed with the economy of the affair and the fact that tangible results were so neatly obtained.

The Mohole Project, apparently going forward at the same time, was running into trouble in the councils of government. The first AMSOC estimates gave a cost of

$17 million. Later the Texas engineering firm of Brown and Root estimated that the Mohole affair could be handled for $55 million. Then someone pointed out that this engineering estimate made no allowance for three years of operating costs. When this item was added, a new figure of $112 million was announced, and the Mohole men now said that it might take eight and a half years to complete the job. Some congressmen were less than enchanted, and Mohole was deep in trouble. Critics of the contract award claimed that it had gone to Brown and Root as a political reward because the Texas owners were cronies of President Johnson and had contributed to his last campaign. (Brown and Root had already performed several billion dollars worth of work for the Defense Department without this charge being made.) The cost estimates for Mohole had advanced so rapidly that suspicious senators and representatives wondered when and if a limit would ever be reached. In the opinion of some observers, the first technical ideas for the Mohole platform would have been a disaster and, in addition, they believed that some of its advocates oversold something to which Congress was already committed. Too much promotion offended some influential men and made them wonder why a thing that was going well needed additional razzle-dazzle.

Finally, the Mohole began to divide the scientists into very polite but very real factions. The dilemma of trying to do two things at the same time, penetrate the mantle and, for good measure, investigate the sediments along the way (get two for the price of one), had led some of the early backers toward a separate drilling program for sediments. The fact that this program was actually accomplishing things (nothing succeeds like success) attracted more scientists to active participation in it. When the attacks mounted against what seemed to many a great and wasteful boondoggle, the defenders found themselves too weak. The Mohole, such a bright and shining dream not long before, was officially killed by Act of Congress of August 26, 1966.

Archie McLerran, an engineering consultant with many years experience in oil drilling, hired in 1964 to work on Mohole, was chosen in 1966 to close out the accounts, saving what he might from the wreckage. At the termination, Mohole contracts already signed, or at least firmly committed, amounted to about $60 million. McLerran worked for 18 months at terminating the experiment. When his labor was completed, the Mohole Project had cost the federal government between $25 and $26 million. (McLerran was able to retrieve a number of things for later use by the *Glomar Challenger*.) McLerran believes that by the time the Mohole was terminated most of its technical problems had already been solved and that the final estimate of $112 million was realistic, final, and could have been met.

Even while the death sentence for Mohole was being written, the JOIDES group had not been inactive. The Scripps Institution, acting for JOIDES, signed the prime contract with the NSF for a DSDP on June 24, 1966. The University of Washington joined the group in 1968. The operation was budgeted for $12.6 million to be expended over a period of 18 months. Anticipating the story, it may be said here that during a time of considerable inflation, the DSDP kept astonishingly close to its financial allowance. Probably the Mohole example made the administrators extremely cost conscious. In contrast to the space program, the basic technology for deep drilling did exist. It was not, however, simply a matter of waving the magic wand and having the results come in. You may know that something works beautifully on paper. It is another matter to move bulky yet delicate pieces of metal miles underneath you down in the cold, black, salty sea. And below all that water lies the immense mass of sediment whose surface has only been scratched and whose depths have only been seen by remote sensors. Everyone with experience knew that the job was perfectly possible. Now all they needed was to prove it.

5

"Our major problem was making people believe the idea was serious."

—Archie McLerran,
National Science Foundation

THE COLLAPSE of Mohole left many people disenchanted, particularly those who are sometimes known as "hardheaded businessmen." The business community was not particularly against pure science, as such, and a number of executives had looked forward to selling equipment to the big project, but what seemed to be mismanagement left them skeptical about any new, government-financed efforts in deep drilling.

The keel of the *Glomar Challenger* was laid at the Levington Shipbuilding Company yards in Orange, Texas, on October 18, 1967, but it might just as well have been merely one more drilling ship. It was the contract that the NSF signed one month later with Global Marine, Inc., of Los Angeles to do actual drilling work that carried conviction. Here was a company, not a federal agency or a university, and one with experience, with whom they could talk. Global Marine developed out of the group that had been responsible for the *CUSS I* that carried out the Mohole tests off Guadalupe Island six years before. *CUSS I* was the joint effort of Continental, Union, Shell, and Superior, and now some of those involved had set up their independent organization. The *Challenger* was the fourteenth ship in the Global Marine fleet. The DSDP had not chosen amateurs. Less than a year after keel-laying the ship completed its acceptance trials and went to work immediately.

When Arthur E. Maxwell, who had been an AMSOC committee member and became Co-Chief Scientist on the *Glomar Challenger's* Leg III, spoke before a group of oceanographers in Paris, he described it as "a rather ungainly looking ship with a very tall superstructure. When it left New York in September I am told that a number of persons who were riding atop the derrick had to duck as they passed under the Verrazano Bridge. Of course, this wasn't necessary as they had at least ten feet to spare —at low tide. Secondly, I should also tell you that after 55 days at sea aboard her, not seeing any land and only one other ship during this time, she eventually transformed from an ungainly ship to a rather handsome one that was extremely effective and could be thought of only affectionately as home."

One rather silly magazine article reported that "the *Glomar Challenger* had a drilling rig put on it." In fact, the whole ship was built around the drilling rig—the only reason for its existence. As a drilling rig, the *Challenger* is a triumph of the latest technology. The living quarters, which are air-conditioned, surpass almost any you could find in oil fields around the world. As a ship, it has few equals for comfort and stays as still as a billiard table with only the barest roll or pitch in the roughest seas. The sumptuous meals served have become almost legendary. Many other people besides Dr. Maxwell agree that the *Challenger* is handsome. It is a new species, bred out of two traditions—the business of the sea and the business of oil.

The business of the sea is a well-known, ancient story, but the oil business, though still young, has also developed a considerable lore and an almost impenetrable vocabulary.

In many places throughout the world, underground oil seeps naturally to the surface. Petroleum, in the form of asphalt floats to the surface of the Dead Sea. In fact, the Egyptians harvested it there for use in the mummification of their dead. Noah used asphalt to caulk his ark. The Babylonians used asphalt for street paving and in the mortar for their great buildings. Marco Polo, traveling in the area now known as the Baku oil fields of Russia, reported that the natives burned oil that came out of the ground in their lamps. The Romans used it for medicine and as a pesticide. The Byzantines employed petroleum for their "Greek Fire," a combination with quick-lime that was the first explosive weapon of the Western World. Two thousand years ago, the Chinese began drilling for oil and for natural gas for light and heat, using bamboo poles for pipes and brass for the drilling bit, but the West knew nothing of this and continued to rely on seepage or the shovel until 1859.

It was then a man named Edwin Drake who realized that salt-well drilling equipment, used for years to drill brine wells and artesian wells, would be perfectly adaptable for drilling into the oil that everyone around Titusville, Pennsylvania, knew lay under the ground. Using salt-well equipment, Drake went down to 69½ feet and then closed on Saturday for the weekend. The next day, Sunday, his partner dropped in to look at the well and found it had filled to the top. An American had struck oil by drilling! Like many developments, it seems so very obvious and simple, after the fact. Should Drake be praised for his ingenuity or everyone else in the region who knew there was oil below simply be called stupid? In any case, Drake's brilliant idea, which may have been thought of a number of times before, caught the world at a moment when it was ready for it. A boom began immediately and within five years Pennsylvania oil became the United States' sixth most valuable export. John D. Rockefeller went into business four years after Drake did and the incredible story of oil wealth began. What Drake's little discovery that oil could be *drilled* did to the pattern of modern life in just a bit more than 100 years almost passes beyond belief. Oil revolutionized not only transportation, and therefore commerce, war, agriculture, cities, and the state of nature, to mention just a few things, but also, to be a bit more particular, it drastically changed the little world of sciences such as chemistry and geology. To return to cases, the *Challenger* could not exist without oil to fuel it and run its generators for electricity.

The amount of oil needed to do all this could not be supplied for long using Edwin Drake's simple drilling technique. The basics, naturally, remain unchanged. A cutting tool is required to make a hole in the earth down to the level where the oil lies, and this tool—the drill— requires power to turn it. As it cuts, the tool moves downward and some way has to be found to transfer the source of power at the surface down to it. The power is ordinarily transferred through the drill pipe and so, of course, this drill pipe has to be strong enough so that it won't break. (The *Challenger* is fitted with very strong, high-tension steel pipe, five inches in diameter, that is considered to be most resistant to the corrosive effect of sea water.) The same pipe through which power is transmitted to the drilling bit brings the oil or gas, if it is discovered, back to the surface.

The pipe is in long sections and these are clumsy to maneuver and to fit perfectly together, as one section after another descends into the hole. Later, when operations are completed, the whole assembly will have to be hauled out. To handle all this clumsy pipe, a derrick is always erected above the hole. Most simply, the derrick supports a block and tackle that lifts the pipe up and down. Derricks used to be constructed of wood, put together with hammer and nails, and during the great oil bonanza on Telegraph Hill in Long Beach, California, in 1921, wood for derricks suddenly became very rare and fetched enormous prices. Now oil derricks are made of steel that can be transported to various parts of a field or hundreds of miles overland to a new project. For traveling derricks, the one aboard the *Challenger* must certainly hold the world's long-distance record.

Oil drilling today has reached the point where its annual volume of business amounts to about $1 billion. The average well now costs around $15 per foot of hole and obviously a 10,000-foot well is a serious affair, particularly when only one well in every eight drilled proves itself a commercial one. Oil-well drillers are often independent contractors who will bid a certain amount of money for a well whose depth is agreed upon in advance. Any unexpected troubles or delays come out of the contractor's pocket. Time becomes money under this arrangement, and so skilled drilling teams are prized for their speed and ability to handle crises. This zest for speed had to be curbed aboard the *Challenger* when too much hurrying turned out to be dangerous.

The earthy crust through which the hole will have to be drilled varies from soft mud to hard rocks such as granite or chert. A drilling bit that is perfect for one type of formation may be inefficient or absolutely useless for another. Scores of different bits have been developed for turning, cutting, chipping, grinding, or hammering away at the formation. An oil well drilled into the soft "gumbo" of the Gulf Coast may use a "fishtail" bit, one with long, widely spaced teeth. Drilling is rapid in such a situation and the teeth are designed to tear the formation so that the cutting will not jam and get in the way. For hard formations, the teeth are stubby and designed to have a maximum crushing or rolling motion on the bottom for the deepest penetration. Very hard bits made of tungsten carbide are often used in tough formations that would quickly wear out an ordinary bit. For the hardest formations, like chert, a diamond drilling bit is sometimes used. This may have as many as 900 carats of industrial diamonds in it. Each diamond costs about ten dollars. A tough formation can destroy a diamond bit in a few hours.

To "make hole," three things must go on all at the same time. The drill string and the bit must rotate. The bit must be lowered continuously as the formation is cut out from under it. And the cuttings produced must be cleared away so they do not interfere with the working of the bit.

The lowering of the whole works, "running in the drill string," is the responsibility of the driller, the man who is boss of the rig and crew. Wearing a hard hat, he stands about twenty feet away from the rotary table, comparatively safe from the heavy pipe that is swinging around. At his side are a number of gauges, the most important of which tells him how much weight is being suspended; and from his gauges the expert driller can tell how the drilling is progressing. With his hands and feet he manipulates a number of buttons and pedals and a sensitive fingertip switch that "feeds off" the brake when necessary.

The whole assembly may weigh hundreds of thousands of pounds but the driller can brake it to within inches.

In drilling for oil, cores of the formation in which the drill is working are only taken occasionally because coring is time consuming and often not necessary. Yet coring is the entire point of the *Challenger,* its reason for being built and going to sea.

The steps leading up to its coring, however, do not differ very much from what would be done on any floating rig when the operators were searching for oil. The *Challenger* was built with a hole amidships through which all the drill string passes. This cannot be seen from the covered-over drilling platform, but below, on the main deck, you can look down into the center of the ship and watch the sea lapping gently about the works. This eerie place is called the "moon pool," a name said to have come from the roughnecks who worked drilling rigs in the famous Santa Barbara Channel.

When operations are to begin, first the drill bit, tungsten carbide or diamond, is fitted onto the bumper sub and drill collar. All this is lowered through the kelly and held for the first section of regular drill pipe. This comes off the automatic pipe racker and is hoisted into position by the traveling block, the machinery that rides up and down the derrick. The automatic pipe racker aboard the *Challenger* was developed some years ago by Archie McLerran, the engineer mentioned earlier who closed out the Mohole operation. The sections, technically called "stands," are three lengths of drill pipe that are fitted together in advance and in ordinary circumstances remain joined. When the 94 foot section is properly over the kelly, the driller lowers it and the roughnecks take hold with great tongs, rather like wrenches, that screw it in tight to the section underneath.

The tongs are pulled by wires from winches off to one side. The very experienced crews of the *Challenger* can add section after section at the rate of one per minute. Even with this speed, it takes a long time to assemble enough pipe to reach down to the bottom, perhaps 10,000 feet below. Enough pipe to do this weighs about 300,000 pounds.

When the driller sees by his gauges that the drill bit should be on the bottom (enough pipe has been let out so that the elusive goal ought to have been reached), he begins to "wash in" the hole (sweeping away sediment with a hose) if the bottom surface is very soft. For this, and for some other operations, several roughnecks have to go aloft. Some go up by elevators that ride up and down the side of the derrick but others are hoisted by lines and they fly gracefully through the air like trapeze artists in a circus or weightless men in space. Once up there, the roughnecks attach hoses to the standing pipe section, hoses that will force sea water down the whole assembly under 500 pounds of pressure. The water washes away the top hundred or more feet of sediment that is too soft for the tough drill to bother with.

Then straight drilling begins on the *Challenger* until the scientists believe enough sediment has been penetrated to make the cores interesting. Now the center of the bit is pulled up and out and the pipe and the core barrel are lowered into this hole in the bit. The metal core barrel is 30 feet long and 2½ inches in diameter. It has a plastic tube inside it and the whole works fits inside the drill pipe and is forced down to the bit by the pressure of the sea-water drilling fluid. The bit now has a hole in the middle of it and, as the bit rotates and works down, the sediment comes through the hole into and up the core liner, like the dough in the center goes into the doughnut cutter. When the liner is presumed to be full, the drill operator hoists it up with a "sand line," which is nothing more than a wire line attached to a winch that winds it up. At the surface the sediment in the core is forced out of the tube by an air pump, and as the round mass emerges from the liner it is cut off in five-foot sections and hurried off to the core lab.

The round trip of each 30 foot core liner takes about two hours in water 15,000 feet deep. When as much as two thousand feet of core may be taken from one drilling site, the operation is long and strenuous. But, to put things into perspective, it may have required more than 100 million years to lay the sediment down in the first place.

6

"The oceanographer has been dreaming of this venture for over ten years and, after a number of false starts, we now have a community effort that is functioning and has brought us successfully to this advanced point."

—Dr. William Nierenberg,
Director of the Deep Sea
Drilling Project

THE *Glomar Challenger* grew from an idea into a ship, symbolically enough, in the yards at Orange, Texas, a small city flush on the Louisiana border and right in the heart of America's southwest. Oil country. When the ship sailed from Orange on July 20, 1968, on tough trials to see whether the men from the DSDP would accept her, it entered the Gulf of Mexico with the most beautiful outfittings that either experienced oilmen or ocean-wise scientists could dream of. Mel Peterson and Terry Edgar, scientists from Scripps, were aboard, but Kenneth E. Brunot, Project Manager, was responsible for deciding whether or not the *Challenger* was, as the Navy says, "ready in all respects for sea." He accepted the ship officially on August 11, after more than two weeks of rigorous testing. It turned out to be almost as good in reality as on paper.

At 10,500 tons, it is bigger than any ship that oceanographers could call their own: 400 feet long, 65 feet wide, and the keel 20 feet below the water line. The ship looks top-heavy to most people. Plunk in the middle a derrick stands as high as an 18-story building, a derrick strong enough to lift a million pounds of dead weight. Most of the forward part of the open deck is buried under 23,000 feet of drill pipe laid out in sections. Another three miles of pipe lies in reserve in the hold.

The *Challenger* carries food and fuel for ninety days at sea without restocking. At two people to a cabin, the ship has room for seventy. Since it has to travel considerable distances between some drilling sites, it can cruise at 12 knots. (The platform designed for the Mohole would have had to get around by use of tugs.) The *Challenger* has twelve diesel-electric engines, all of the same design. Some ships carry various different kinds of engines and this creates a very serious inventory problem when it comes to storing spare parts. The planners had bought two of almost everything the ship might need. The engines worked so well that during the first 18 months of almost continuous operation, the ship only lost 124 hours due to their breakdown.

The engines are part of the ship's unusual capacity for "dynamic positioning." This is a wonderful system for keeping the ship exactly where you want it, no matter what the winds, or the sea, or the ocean currents may be, without benefit of any anchors or lines or any other outside force.

When the scientists decide that the ship is where they want it to be, a red sonar beacon is dropped over the side and sinks to the bottom and then begins sending signals back to four hydrophones beneath the ship. These receivers feed their messages into a computer that has been told where the ship should be. Knowing from its relation to the beacon where it actually is, the computer figures out how the ship must be moved to put it on the location. It sends commands to the main screws and four "side thrusters," two on the bow and two on the stern, that can move the ship sideways if necessary. The more drill pipe it has out, the more off center the ship can be, but dynamic positioning usually keeps it well within this range. The side thrusters, which very few ships have, usually startle and amaze the people who always seem to be standing around a pier when a ship docks. They are unable to understand how a ship can just move sideways, without benefit of tugs.

The computer controls are situated prominently on the ship's bridge. Computers can be fallible. At first, during the tests, dynamic positioning worked very well and then, after about a half hour's operation, the ship would suddenly go "Full Speed Ahead," no matter how much pipe it may have had out. This illness was finally diagnosed as program failure and the man who put the program into the machine was fired without one kind word.

If necessary, dynamic positioning can be done manually. As the computer makes adjustments to keep in position, its orders to the engines are recorded and a pattern for the orders usually emerges. Breakdowns of the computer have occurred and the mate on watch has been required to take over. To match the speed of the computer and to reckon with all the variables is a severe strain and the mates have only been able to do this work for about 15 minutes before they needed to be relieved. Nevertheless, the ship has sometimes been positioned manually for a number of hours. When the computer breaks down, it takes two and one-half hours to put a new program into it.

If the beacon on the bottom should fail, the mate on watch quickly puts an average of its last two hours of signals into the machine. Should the failure continue for any length of time, another beacon is dropped overboard, one that sends on an entirely different frequency. The first beacon might suddenly come to life later and, with similar but different messages, render the computer and the ship completely helpless for position keeping.

Should an emergency occur, with miles of pipe hanging below, where it was necessary for the ship to get underway immediately—such as being approached on a collision course by another ship that was completely out of control —the *Challenger* could move even with all its impediments, but everything below the keel would probably be ruined and the ship would be almost impossible to maneuver. Such a situation is extremely unlikely but it has been thought about, and the decision was made that in such a case a charge would be sent down and detonated to blow up the pipe.

One of the two men who are in command of the ship during alternate cruises, Captain Joseph A. Clarke, had had a sign put up on the bridge that reads, "A collision at sea can ruin your whole day. Thucydides, 450 B.C."

The *Glomar Challenger* is remarkably stable in every kind of weather due to a gyroscopically controlled system that keeps rolling and pitching down to a minimum that is hard to believe. It is the first civilian ship in the world to receive weather photos from a satellite. Each day when the Navy's weather satellite is in proper position, a man from the Weather Bureau on board the ship adjusts his instruments to receive an up-to-date satellite photograph of conditions in the sea around him. The rapid forecast is of great value in planning drilling operations and giving warning of severe storms. The ship also receives a facsimile weather picture every day by radio.

The *Challenger* is also the first commercial ship in the

world to use the once-classified satellite navigation system developed by the United States government. For oceanographic research, it is extremely important to know exactly where your observations are being made and particularly so when the matter under consideration is a hole one-half mile deep in the earth. Older navigation systems might give the ship's position with an error of a mile or more. The satellite is supposed to be accurate within .1 of a mile, but in practice it can be correct to within several hundred feet. As a timepiece, the satellite orbiting 500 miles above the earth is correct to within 30 milliseconds. It moves at 17,000 miles an hour and its orbit seems to be almost perfect. There is a satellite passage about once every two hours and the times of passing are reckoned by the computer that is part of the equipment. The computer aboard the *Challenger* takes six minutes to fix the ship's position and this is much too slow because in those six minutes the ship can drift substantially away from the position at the time the reading was taken. However, once *on* a drilling site, a great number of fixes may be taken on the satellite (at one site the operator took 64) and a very good average position can finally be established.

This, then, was the ship and the special equipment that the participants for Leg I took over immediately after the acceptance trials. Robert A. Wilson was Captain for the first voyage. Dr. Maurice Ewing, the tall Texan with the leonine head of hair, was Co-Chief Scientist along with Dr. J. Lamar Worzel, Associate Director at Lamont and second-in-command to Dr. Ewing there. Dr. Worzel had been at sea on 34 oceanographic expeditions and been Chief Scientist on all or part of 26 of these. He had done a great deal of work on the sediments and structural history of the Gulf of Mexico.

Among others on board were geologists from Mobil Oil and Continental Oil; Dr. Alfred Fischer, a geologist from Princeton; paleontologists from Massachusetts, Texas, and California; Darrell Sims, the DSDP engineer; Archie McLerran, who was now officially the NSF representative to the Project; and a number of technical people, including eight from Scripps. Truly a cross section of the United States.

The first hole on the $12.6 million Project was drilled August 12, 1968, the day after the acceptance trials. Various delays has already put Leg I a week behind schedule, and the two Chief Scientists wanted to waste no time. The first site was in the Gulf of Mexico on the Sigsbee Abyssal Plain, just a few miles from a steep escarpment that separates the truly deep ocean from the continental shelf. Its position was 21° 51′ N, 92° 11′ W, roughly midway between Louisiana and Mexico's Yucatan Peninsula. In terms of underwater topography, this is on the edge of a basin known as the Sigsbee Deep, in honor of the American who first discovered it. The site was drilled mainly for practice in deep drilling and coring so the site was in water nearly 10,000 feet deep. Over 12,000 feet of drill was finally let out and the last core liner into the hole went 2,528 feet into the sediment. (Before Leg I was completed, drilling would be done in 17,567 feet of water, using 18,523 feet of drill pipe on one occasion but the best penetration would be the first hole.) The oldest sediments were merely Pleistocene, the last two million years, with few fossils, mostly a carbon-rich clay from turbidity currents, somewhat similar to landslides, that had flooded the region during the Ice Ages.

The second and third sites were the great objectives for Dr. Ewing in the Gulf of Mexico and he was not disappointed in the drilling results. These were in the region of the Sigsbee Knolls and he had discovered the first three of these knolls in 1954. Later surveys showed that the flat, abyssal plain had 44 diapirs or domes on it, all buried beneath the sediment except for Ewing's original three. Ewing believed that they were all salt domes for a number of reasons, among them the facts that they were intrusive, but not small volcanic peaks, as shown by gravity and magnetic surveys, and they were neatly rounded. They also closely resembled the salt domes or small hills found on the continental shelf of the Gulf of Mexico. This region has been a deep ocean for an extremely long time and, as Dr. Ewing explains in a statement about Leg I, "the suggestion that these all were salt domes was strongly opposed on the grounds that extensive, thick deposits of salt, of the type from which salt domes are known to grow, could not conceivably have been deposited in a permanently deep basin."

On a cruise in the early part of 1967, the Lamont ship *Vema* had outlined the zone of the knolls and domes and showed that they joined the known salt dome fields of Tabasco-Campeche, an oil region of Mexico. The particular dome that was drilled was chosen because it broke above the ocean floor and could be located, therefore, by satellite navigation. Later Dr. Ewing and Dr. Worzel named it the Challenger Knoll.

At the second site, the water was 11,753 feet deep and the drill bit penetrated 480 feet into the salt dome on the floor of the Gulf of Mexico. Salt domes are masses of salt that have grown by thrusting through thick beds of sediment. They are topped by formations known as caprock, typically composed of lime, sulfur, anhydrite, and gypsum. Oil and gas often accumulate in this caprock. The bottom core, number 6, penetrated the main mass of caprock on the knoll, rock that consisted of gypsum crystals veined with residual oil. Dr. Ewing said, "It smelled just like East Texas in the boom days."

The significance of Site 2 is that it showed that oil (hydrocarbons) could be formed and accumulated under deep-sea conditions. For many years, scientists have been trying to solve the mystery of how oil is formed and the general belief was that it had to happen in shallow seas. This find gives the geologists a new problem to handle. As written in the official report, "This hole was abandoned and plugged with cement to prevent seepage since there were rigid instructions to avoid any possibility of an uncontrolled flow of oil." Dr. Ewing would have liked to drill far deeper in an attempt to find more about the age of the salt beds.

Looming as large as the scientific question raised by Site 2 is the economic one. The *Challenger* found, for the first time in history, that oil lies far out in the deep ocean. Is this unusual, or does it exist in many places? Is there enough oil to make a fully producing well? How will it be recovered from a depth of 12,000 feet? In principle, production from this depth should be possible since the same kind of job is already being done farther down than a diver can swim and, below this, all technical problems are essentially the same.

The Sigsbee Knolls are on no nation's continental shelf. If they are really a potentially valuable oil field, whom do they belong to? Would the United States and Mexico or Cuba be willing to share them? What about the Malta Resolution at the United Nations, which suggests that deep-sea wealth be shared by all the world? The United States petroleum industry, in the year of the *Challenger* discovery, recovered $4.5 billion worth of oil from shallow, offshore fields. The deepest wells were about 1,000 feet below the surface. What kind of a genie has the DSDP unbottled?

Several months before the *Challenger* sailed to the Sigsbee Knolls, some alarmed people had warned that if there were oil at high pressure within the domes they might blow out the drill pipe with explosive force and wreck the whole drilling rig. The Project experts replied by pointing out that the highest pressures ever found in a well so far were 8,000 pounds per square inch and that the water pressure in the depth they planned to drill would be 6,000 pounds per square inch, enough to subdue any possible explosion. Nevertheless, precautions were taken that frustrated the scientists.

Some of the sediments from Site 2 were given to the American Petroleum Institute for study and they distributed them to a number of oil companies with the understanding that there would be no secrets permitted and all findings would be published. The Petroleum Institute reported later that the oil "is relatively young, that the rock is mainly calcite and sulfur and that the rock contains an accumulation of fossil pollen (grains) of Jurassic age."

Site 3 was drilled near the Challenger Knoll to take samples of normal basin sediments in the vicinity and to see what kind of material beneath the surface was acting as reflectors on the seismic profiles. The penetration reached down into Miocene sediments, perhaps 30 million years old, but then was abandoned because the explorers had received an order that the maximum penetration permitted was 2,000 feet and they had reached 60 feet farther. As the *Challenger* sailed on to Site 4, Dr. Ewing anticipated that the future debate will be whether the salt for the Sigsbee Knolls was deposited in the present deep water or deposited at some earlier time when the crust may have been much shallower than at present.

From the Sigsbee Knolls in the Gulf of Mexico, the *Challenger* sailed more than 1,000 miles east before making the next stop on its pioneer cruise. The scientific destination was in the Bahamas area, about 60 miles northeast of San Salvador, the little island where Columbus may or may not have made his first American landfall. Ewing and Worzel had chosen this part of the ocean floor because they had found, in surveys during the previous ten years, that bottom reflectors observed over most of the western Atlantic outcropped through the sediment here. Seismic profilers working all over the ocean had found a characteristic signature beneath the top sediment and this was named Horizon A. Underlying Horizon A, but not found so extensively, was a somewhat similar reflector the discoverers called Horizon Beta. Another reflector, quite a bit deeper, was given the name Horizon B. Horizon A had been sampled northeast of San Salvador by Lamont ships using piston corers, and the sediments had ranged back to early Cretaceous in age, perhaps 135 million years old.

Here at Site 4, Horizon Beta was found on the profiler to be about 50 feet beneath Horizon A and Horizon B, about 1,200 to 1,400 feet farther down. The goal was to take samples of all three horizons if possible.

A hole was drilled where the water was 17,452 feet deep. The roller bit was destroyed by chert after it had penetrated 849 feet. A total of 48 feet of core was recovered. Since Horizon B had not been reached, another hole was tried with another kind of bit at the same position. 600 feet below the bottom, this bit also encountered the extremely abrasive chert. Then, after working down another 80 feet, the core barrel jammed and operations had to cease. When the pipe was pulled to the surface, it was seen that the bit had been absolutely destroyed while trying to work through the chert. Also, it was discovered that the profiling gear would not work while the ship was at this location.

Leg I had already had its share of hard luck. Acceptance trials had taken longer than planned, one young man with a badly broken hand had to be taken off the ship by the Coast Guard, another man removed with a severe kidney-stone condition, and then one of the roughnecks on the drilling crew was struck in the chest by the heavy tongs that tighten the drill pipes together.

To observers it looked as if the blow had actually fractured his heart. His chest was hemorrhaging badly and no one thought he would live as the ship hurried him to San Salvador. From San Salvador, he was carried to a hospital in Nassau, the capital of the Bahamas. X-rays showed the man's ribs were not even broken. After all, he had merely sustained a massive bruise and later he rejoined the ship.

Another human crisis occurred when Dr. Alfred Fischer, the scientist from Princeton, disappeared. During the afternoon someone asked where he was and no one seemed to know, but the *Challenger* is a big ship and he could have been almost anywhere. Then when he was really needed to make a decision, it was realized that no one had seen him for more than three hours. An organized search began but Dr. Fischer simply could not be found. The captain was about to call for "man overboard" procedures when the geologist reappeared. It was a pleasant, sunny afternoon and he had simply climbed to the top of the derrick to read a book. Professor Fischer did not understand, at first, what all the fuss was about.

It was decided to abandon Site 4 to the chert and try another likely part of the ocean floor 20 miles to the northeast. The new site chosen was a place where the reflector could be much nearer to the surface and, therefore, the cherts either much thinner or they might even have outcropped to the east, giving no trouble at all.

Unfortunately, the *Challenger's* profiling gear still would not function so a new location had to be found on the basis of satellite navigation and position depth records taken previously. Site 5 was somewhat farther to the east than had been planned and the topmost section thicker than desired. Then mechanical difficulties with the drill string, one of the causes for abandoning the last hole, began again, and the assembly had to be recovered to clear up the difficulty. A new hole was now drilled at the same spot, using the same diamond bit. This time a penetration of 914 feet was achieved. Profiler records indicated that there was much farther to go before the reflector could be reached, but no more progress could be made. When the bit was recovered to find the reason why, it was seen that the diamond-cutting tool had been completely destroyed.

Then the *Challenger* received a radio message that New York, the next destination, was threatened with a dock strike and the ship had better arrive early if it planned to take on stores and change crews before the strike closed operations down completely. A delay at New York could upset plans that had been made at least a year ahead. To the distress of the scientists, no further drilling was attempted in the Bahama area.

Yet, on looking at the little bit of sediment taken on the very last core, the paleontologists found that the fossils in the sediment belonged to the Upper Jurassic period, the age of dinosaurs, about 150 million years ago.

This sediment was much older than any that had ever

been taken from the deep sea. The *Challenger* had another record, while still on Leg I.

With the Upper Jurassic core, time had been pushed back and the Atlantic was suddenly 150 million years old, probably even older. The resistance to the new geophysics, unpopular with some older scientists, soon flared up after this discovery.

"In this area it was found that deep sea conditions much like those found today prevailed here for at least the last 150,000,000 years, that beneath these are much older sediments which place serious limitations on theories of Continental Drift by sea floor spreading. . . ."

Other scientists commented that finding sediments of this age, or even 50 million years older, had no real bearing on the idea of continental drift. The finding merely showed that the Atlantic opened up some time sooner than had been previously thought.

Whatever was proven about drift, the work did demonstrate that the *Challenger* performed very well at keeping position in extremely deep water and that the longest strings of drill pipe ever used in the deep sea would hold under the quite considerable strain.

In terms of further *Challenger* operations, the unexpected chert problems were the most important discoveries of the operations in the Bahamas. The Worzel-Ewing report stated: "These layers may be sufficiently widespread in occurrence to present a serious problem to any program of deep-sea sediment sampling. The oldest sediments may remain measurable to the seismologist but beyond the reach of the geologist until our technology provides better bits or a means of reentry to a hole in several miles of water."

Although it has been known to geologists for a long

time, chert continues to baffle scientists who wonder what causes it. A scientific encyclopedia says, "A great deal has been written on the occurrence and origin of chert and there is no doubt but that it may be formed in several different ways."

Dr. Alfred Fischer, a geologist on Leg I and an authority on the Alps, pointed out that marine sediments, such as chert, that are found as rocks high in the mountains were thought to have become cemented after they had been lifted out of the sea. Leg I showed that the cementing could have occurred under the water. Chert, dark and hard as flint, may be formed by some process out of the ocean beds of turbidite, the masses of debris that sometimes flow along the ocean floor like landslides. These turbidities mix with the dissolved shells of organisms, such as radiolaria, that are rich in silica and then the whole mass recrystallizes to form the flintlike chert.

A course was set for the *Challenger* to a position on the Bermuda Rise, south and west of the island. Here the ocean floor slowly rises up above the surrounding abyssal plains toward the central Bermuda Pedestal where the feature breaks the surface. The Bermuda Rise is a broad swell of the earth, more or less in the shape of an ellipse, more than a thousand miles long from southwest to northeast and about six hundred miles wide. The northeast half is heavily sedimented, probably from the ranges of the Kevin and Muir seamounts that lie north of it. The southwest area, where the *Challenger* went, has a relatively thin sediment cover and it was thought that a complete sequence of sediment to the rock basement might be acquired. It was also believed that the western flank of the rise was elevated enough above the plain so that the drillers could avoid the turbidities associated with chert. An earlier track of Lamont's *Vema* (the

Challenger was carrying some of its profile records) seemed to show a location with a good sediment sequence and the presence of Horizons A and Beta.

At the site on the Bermuda Rise, the *Challenger's* profiler became "marginally operative" and an unexpected peak was observed rising above the basement. The position *Vema* reported must have been in error, since it did not have a satellite navigation system at the time it made the survey. The scientists decided to hold their course and speed until the *Challenger's* own profiler showed a proper site. Site 6 was established five miles beyond the submerged peak no one wanted. Horizons A and Beta lay fairly horizontally here and not too far down for drilling. At this station, too, it was thought possible that important animal fossil changes might be documented at the boundary between the Mesozoic and Cenozoic eras, two of the greatest stages in earth history, through which the drill should pass.

The bit went down at Site 6 through 16,815 feet of water and then into 840 feet of sediment where six cores were taken. Then drilling progress stopped and the obvious conclusion was that chert had again destroyed the bit.

A hard layer had been reported by the drillers for the first 20 feet after the first contact was made with the bottom, but the scientists had no sample of it. As soon as the bit was clear of the bottom, the drill string was lowered several times in attempt to sample this hard surface layer. This time no hard surface layer was found, only soft, brown deep-sea clay, so no explanation was found in this extra punching operation.

Since Site 6 had proven to be a thick series of chert layers and only a short time was available, the decision was made to sail the ship about 50 miles farther south-

west where seismic reflections had said the sediment between Horizons A and Beta was much thinner. With only
a few hours remaining on this first leg of the *Challenger*
cruise, a sample would be taken of the sediment underneath Horizon A. Thirty-two feet of brown clay were
taken in two cores from the first hole and then a severe
swell (to which the ship was allowed to remain broad-
side) caused the drill string to be bounced on the bottom
of the hole, damaging the core barrel. So the drill string
had to be brought back 17,000 feet to the surface, taking
precious time, and the core barrel replaced. A second
hole was dug and coring began after the bit had penetrated 872 feet below the bottom. Cherts of Horizon A,
probably mid-Eocene, about 50 million years old, were
finally penetrated after a long effort and clays of early
Eocene or Paleocene were cored. The last cores were
about 15 feet long, without fossils, and the bottom represented the early Eocene or Paleocene age, between 60
and 75 million years old, or perhaps somewhat older.

"At this point in time," said the Initial Report, "the
ship's navigator judged that departure for port was necessary in order to have an adequate margin for maintaining
a necessarily inflexible schedule. The hole had to be
abandoned at this point."

No one needs to have been on the *Challenger* bridge
right then to guess that relations were very difficult between Dr. Ewing, a certified grand old man of oceanography, and the ship's operators who believed that their
responsibility was to get the *Challenger* into port when
it was due. Was it more important to keep drilling, once
the ship had been taken out there after an enormous expense of time and money, or was it more essential, as the
administrators on shore certainly felt, to keep a timetable
planned more than a year in advance that involved very

complicated logistics and concerned the plans of hundreds of people? Not taken into account, but certainly important, was the fact that the crew had already been at sea for seventy days. Whoever was Chief Scientist on the maiden, exploratory voyage would probably have been in the same situation.

Of course, whatever tempers were displayed or kept in tight rein, Leg I had been a success. The *Challenger* came into New York on September 23, 1968 (or rather to Hoboken, New Jersey), and Dr. Ewing called the program "a new era in the science of geology."

A press conference was held and the work of the ship was praised and explained by major newspapers. Columbia University held a symposium on deep drilling in connection with the awarding of the Vetlesen Prize in geophysics (worth $25,000) to Dr. Francis Burch of Harvard and Sir Edward Bullard of Cambridge. Dr. Ewing had received this same prize a few years before. The symposium speakers talked about the *Challenger* findings and the learned audience was considerably impressed, but it was a meeting of the club, the "in group" of a science that was very high at the moment. The *Challenger* men happily accepted the praises due them, then went to work analyzing and understanding the 575 feet of sediment core they had taken during the long and sometimes frustrating weeks at sea.

7

"It is evident from the results of this pioneering cruise in deep-ocean drilling that the *Glomar Challenger* and its system work well."

—*Initial Reports*, Volume I,
of the Deep Sea Drilling Project

LIFE ABOARD the *Challenger* on Leg II developed a style, an atmosphere that has changed very little since the shakedown days of Leg I. When men live isolated together, with few of the normal patterns of home, interest outside their work often centers on food. Meals are the focus of social life, the break in monotony. The gripes of soldiers and the prison riots most often develop out of discontent with the mess hall. The owners of the ship, Global Marine, very wisely took care that no one would ever find a reason to complain about the food. The quality is so good, the menu so varied, the portions so lavish that meals surpass those on most luxury liners. This is dining, not just eating, and it creates an extraordinary sense of contentment. And the cooks are so cheerful! They are very genuinely concerned that you enjoy your meals and will go to great trouble, if they have the ingredients, to prepare anything special someone might ask for. When Archie McLerran asked for corn bread at lunch, it was available, hot and fresh, for dinner. The only complaint may be about having steaks more than once a day.

Everyone eats together, including the captain, the most honored scientists, and the deck hands. A certain amount of rank does exist on the *Challenger* and the men tend to sit more often with fellows who have the same interests, but there is no enforced stratification of society like that found in the Navy. The *Challenger* occasionally carries two women scientists, and they are treated with courtesy and as equals. There seems to be little of the compulsive use of four-letter words in every sentence that is customary, almost required, in most similar military situations. This is a society of adults. Some of the younger scientists have worn long hair but the crew has been, at least publicly, tolerant of them.

Before the *Challenger* sails for its planned sixty days at sea, the menus are planned and printed. Breakfast is always the same. There is a choice of chilled prunes, apricots, peach halves, pears, and pineapple; in season, canteloupe, honeydew, or grapefruit; also fresh orange juice, tomato, prune, grapefruit, apple, or grape juice. Eggs are prepared in any style. Then there can be bacon, ham, or sausage. Hash brown potatoes are served with all orders. Creamed hamburger or chipped beef as well as corned beef hash are available. There are hot cakes with maple syrup or honey. A variety of fresh breakfast rolls, hot donuts, and hot biscuits is baked early each morning. There is toast. And of course, coffee, tea or milk, and assorted jams and jellies.

The drilling crews work twelve hour watches, one from midnight to noon, the other noon to midnight. When they come off watch they are hungry. A midnight meal is served every day for the men coming off or going on watch and anybody else who wants to eat is also welcome. At this late meal, juices, fruits, and eggs, hamburgers, cheeseburgers, ham and eggs, beef and grilled cheese sandwiches are offered. There are omelets, any style. For those who want to construct their own sandwiches, there is a cold tray with bologna, Italian salami, summer sausage, head cheese, pastrami, pimiento loaf, ham, Swiss and Cheddar cheese. Hot meals are either vegetable soup or chili and beans, Texas style, with soup. Donuts, rolls, biscuits, and homemade light bread. Assorted pickles, mustard, mayonnaise, sliced onions and olives. Lettuce and sliced tomatoes are also available while still fresh the first few weeks out of port.

Breakfast and the midnight meal are, of course, merely sideshows compared to lunch and dinner. Lunch and dinner are hard to distinguish from one another, in the sense that one is quite as imposing as the other. For variety, during the cruise, the following are offered: French onion soup with croutons, avocado halves, asparagus spears au gratin, buttered hominy, broiled lobster with drawn lemon butter, Manhattan clam chowder, shrimp creole, sauteed mushrooms with white wine, iced fresh oysters with dip sauce, pan fried frog legs, roast tom turkey with oyster dressing, Virginia baked ham, beef jambalaya, enchiladas, refried beans, grilled swordfish, beef stroganoff, and chicken cacciatori. These were special and not all of them appeared on every menu.

One quite ordinary dinner suggested: Soup de jour, dinner steaks, roast sirloin of beef, ocean fresh Mahi-mahi, baked potatoes with sour cream and chives, corn on the cob, fresh asparagus, steamed cauliflower, steamed rice, coleslaw, carrot and raisin salad, potato salad, iced celery stalks, boiled eggs, olives, pickles, and peppers. The pies baked specially for dinner were peach pie,

cherry pie, chocolate pie, coconut cream pie, lemon me-
ringue pie, pineapple cream pie, pecan pie, and cherry
Boston cream pie. There was also Jello.

In case anyone got hungry aboard the *Challenger* between meals there was an instant ice cream dispenser,
fresh milk, fruit drink, coffee, hot or ice tea always available. The refrigerator was open to anyone who wanted
one of a variety of soft drinks. For afternoon the cook
also put out, on a typical day, almond macaroon cookies,
glazed chocolate brownies, peanuts, apples, oranges, tan-
gerines, glazed donuts, corn sticks, potato sticks, Dutch
pecan squares, and almond butterfly buns.

Running such a restaurant sounds as if it should be
very expensive, but apparently it is not. The secret seems
to be that the men involved seriously care about what
they are doing. The Chief Steward on one of the alternate crews, Clinton D. Rouse, said, "You've got to have
a cook. Not someone who stays in his sack."

Planning is also a great part of the commissary's success. A whole turkey is served as a roast at one meal.
Then it will be used as hash, or in a Chinese food such
as chow mein, or sliced with cream sauce. All these combinations are planned before sailing. Should any of the
turkey be left, it might be served as cold cuts. But practically nothing goes over the side as garbage. Any left-
overs are *converted* the next day but nothing is kept any
longer. Still waste hardly exists. There is little that can be
done with leftover vegetables and the cooks do not even
try. Such vegetables are not even used in soups. They
have poor eye appeal.

Keeping all this food for several months at sea without
spoiling is a problem that has been very well solved. Surprisingly, fresh milk is always available. The books say
that milk cannot be frozen for keeping, but it can if it

is homogenized. On the *Challenger* it is bought in four-quart plastic containers that are put in the deep freeze the moment they come on board. A day before use, they are transferred to the regular refrigerator for thawing. Butter is also frozen and when served seems absolutely fresh. The Steward has looked for some supplier who could freeze cottage cheese and sour cream but has found no one so far who has figured this out.

For fresh things, lettuce goes first, but it will last for three weeks if kept at a temperature between 34° and 35°F. Flour is kept in a cooler at 35° to 38°F. because every kind of flour has bugs in it and these will begin to develop sooner or later, particularly in the tropics. All starchy foods are kept in a cooler but not soda crackers. Nothing can keep crackers from finally becoming soggy. Perhaps one day even this hardship of the seagoing life will be overcome.

When not eating in the mess hall, some of the crew gather in the mess hall for pinochle games, but mostly for talk and the inevitable coffee that is part of the fuel supply of every ship flying the United States flag. For sleep, the accommodations are two men to each air-conditioned cabin and two cabins share a bathroom with shower. (The ship carries large amounts of fresh water and also evaporates it efficiently so water rationing has hardly ever been necessary.)

The crew's lounge and the science lounge both have movie projectors and anyone who cares to can operate the equipment. The choice of films is not great enough to show a new feature every day and the quality varies considerably. (Global Marine rents the pictures, but whoever is in charge of this operation obviously has other, more important, jobs as well.) Laundry left outside the door in a mesh bag is returned the same day. Washing and drying are done while the clothes remain in the bag so workday clothes get no ironing.

Liquor and gambling are absolutely forbidden on the *Challenger*. Global Marine does not want to repeat the experiences of other drilling ships where men have had losses so great that they were absolutely broke, and even in debt, at the end of a cruise.

The ship has a radio-telephone for official business but a man anxious to talk to his family may sometimes do it by means of a "phone patch." The United States has many amateur radio operators who enjoy getting calls from ships, dialing a specified phone number, and patching a connection from the ship's transmitter, through the ham radio station, to the stateside telephone. The only charge is *that* within the United States, and the calls are all made collect.

One source of recreation is a large gym created out of empty space below the long pipe racks. A small profit made out of the sale of tax-free cigarettes at the ship's store has paid for a Ping-Pong table, darts, exercise horses and bicycles, punching bags, weights, and boxing gloves.

Nature also provides entertainment. The *Challenger* usually has operated in warm seas where fishing may be excellent. Mahi-mahi (a dolphin fish, not the mammal) are often caught and given to the cooks. Sailors on any ship like to catch and kill sharks and the *Challenger* crews have had the same built-in hatred of sharks that is practically universal among men who go to sea. The fish are also very interesting to watch. The *Challenger* is often stopped for long times to drill and many curious animals come around to look. Whales and squid are sometimes attracted to the vibrations coming out of the acoustic beacon and porpoises, of course, often appear. Of all the forms of life seen in the water off the *Challenger* decks, the dolphin fish may be the most beautiful. About two or three feet long, they swim close to the surface, foraging, and their dark blue-green bodies, light blue fins, and forked yellow tails make one regret eating them.

Many land birds find the *Challenger* from time to time. Some may have stayed with the ship after it sailed from port but others that have no business at sea are sometimes driven hundreds of miles from land by storms. A humming bird and a hawk, along with more ordinary birds, have joined the ship and an owl who must have arrived by accident was fed raw meat and water for a month, but the crew's efforts could not save it at the end. One night off the United States East Coast, a flock of small yellow birds, obviously lost in a fog, crashed onto the decks and more than 400 bodies had to be swept overboard in the morning.

The *Challenger* usually has seventy people aboard during a Leg. Forty-five of them are employed by Global Marine and the company has no trouble at all finding men to hire. The crews alternate, one voyage on and then one voyage of vacation. One day's vacation for every day's work is one attraction of the job but, as it actually turns out, the men put in as many hours at sea in six months as another man does on a shore job in a year. The crew are somewhat older than the average for the merchant marine, men with families who like to know months in advance exactly when they will be home. They and their wives can plan well ahead and, for sailors, the men on the *Challenger* seem to have very happy married lives.

Crew morale is excellent because of the good pay,

good officers, and the excellent food, but also, according to Ken Brunot, the Project Manager, because "they have a sense of participation, of doing something that's never been done before."

The least skilled roughnecks earn about $14,000 a year. A driller can make about $20,000. For this they do 12 hours of hard work, often dangerous work, every day including Sunday for 55 to 60 days at a time. Yet Global Marine has a waiting list for jobs.

Men who work on oil drilling crews could not survive routine factory jobs. As John Steinbeck pointed out, they are tough and highly skilled, an elite and quietly proud of the fact. As Ken Brunot has noticed, they are always very interested in new equipment and anxious to see how it works.

The nature of the work makes oil drilling a transient job. Once the well has been dug, you move on. Nature has seen fit to bury oil in many difficult places and being aboard a ship as comfortable as the *Challenger* for two months may seem preferable to drilling Alaska or the deserts of Arabia. The men on the *Challenger* drilling teams have home addresses like Purvis and Brookhaven, Mississippi; Columbia, Herbert, Winnsboro, and West Monroe, Louisiana; Salado, Arkansas; Oklahoma City; Gilmer, Daisetta, and Odessa, Texas; Lynnwood, Bakersfield, and Oakview, California. Most of them have had ocean drilling experience on the shallow water rigs in the Santa Barbara Channel. When the scientists on Leg II held informal classes to talk about the science objectives and accomplishments of the drilling, these crews were the men who all came and returned for more. Learning their trade, the drilling men had always talked in terms of feet, but the scientists were in the habit of using meters when discussing depths. The drillers have tried to go along with the new language, thinking and talking meters, a difficult switch to bring about naturally.

Life for scientists at sea is less muscular than the drillers', but when a fresh core comes aboard it may be just as exciting. In the words of Mel Peterson and Terry Edgar, "For a geologist it is like being let loose for the first time in a library of a civilization lost forever to time." The inanimate material in the long tubes can tell the trained man about the oceans of ancient time and the climates, about the eruptions of volcanoes and when mountains were built, how ocean currents have changed, how the present living animals evolved from the past, and where our present oceans and continents were in the past.

The kinds of minerals in the core suggest what land they derive from or what volcano. Bits of soil tell about ancient climates and old continents. Animal fossils are part of the story of life, their thinness or profusion saying how productive the ocean currents used to be.

Such material tests old theories derived from indirect evidence, proving some, disproving others, and suggesting entirely new ideas to the scientists. Each sample "brings us closer to the ultimate goal of reconstructing the earth's history." Maurice Ewing, Co-Chief Scientist of Leg I said, "To a geologist there is nothing quite like a sample."

When the drillers have brought up the core barrel, two very small samples are taken of the contents, one from each end, and they are smeared on slides. These are quickly looked at under a microscope. This is a fast determination of the approximate age of the sediment and its type. The work is done by a micropaleontologist and then it is up to the Chief Scientist to decide immediately whether to continue coring, drill deeper before coring again, or to abandon the hole.

Then the core is cut into manageable five-foot sections that are brought into the core laboratory, the most complete ever built for geological work at sea. They are weighed to determine their density and X-rayed to analyze their internal structure. The X-rays, among other things, show how the sample may have been disturbed during coring. Measurements are then taken of the core's natural radioactivity and its porosity and density. The velocity of sound of the material in the core is measured and this is compared with the velocity of sound found by echo sounders at the same depth. Thermal conductivity is read and, with a knowledge of the temperature in the hole, a computation can be made of how much heat is flowing out of the earth's floor at this site. All this takes between one and a half to two hours, and since the cores are being brought in at the rate of thirty feet every two hours, the laboratory soon becomes knee deep in mud.

Then each five-foot section is sawed in half and the working half described and studied for its minerals under a petograph microscope. It is also sampled for its working properties. Then it goes below to the micropaleontologists who examine it more carefully under the microscope to see, by the species of fossil they find, the age of the core, and sometimes the ancient climate in which the species lived.

The other half of the core, the archive half, is photographed and sealed away without further disturbance. Finally, both halves of the core are put in a refrigerator and kept at about 35°F. so they will retain their moisture. When coring is going well and more and more samples are coming in, there is pressure on the technicians to speed up their work. Mix-ups, however, such as getting the cores out of sequence and rearranging a few million years of sedimentation would be terrible. Samples from a thousand feet beneath the ocean floor are still very rare. They are, incidentally, carefully guarded and kept away from shipboard souvenir hunters.

On board the ship, the cores are summarized in graphs that, meter by meter, give their geological epoch, magnetic orientation, their radioactivity, length, general rock type and then a more detailed discussion. The first half meter of the core on Hole 1, Leg I, was described as "Clay, rich in coccoliths and globigerinds, brownish-gray, massive (disturbed by coring)."

Next, graphs are prepared in centimeters which show the core's X-ray, its picture in black and white (taken just after it emerges from the sea). More details about it are given in terms of geology and fossils. Then the sediments found are considered in a very general way and their nature given in a general description.

Once Leg I ended, the cores went finally to the DSDP repository at Lamont-Doherty in Palisades, New York. They were subjected to analysis for grain size, water content, carbon carbonate, and paleomagnetism. The relative mineral abundances were studied by X-ray diffraction at the University of California. Various fossils were studied at Scripps, the University of Miami, at British Petroleum, and at Woods Hole. The oil and cap rock were exhaustively analyzed by scientists from the oil companies. Finally, in a 672-page volume describing all these tests, summarizing essays were written on all the sediments, in terms of both sedimentology and paleontology, and on the completed geophysical picture of the whole area as seen at the end of Leg I.

Sediments are considered to be either pelagic, terrigenous or of volcanic origin. The meaning of pelagic has been hotly debated but, practically, it means sediments of the deep, open ocean that settled down from the waters above, far from land. They are either the remnants of some kind of plant or animal or some kind of clay. Terrigenous sediments derive, in several ways, from land.

Many pelagic sediments are brown clay, usually soft, plastic, and greasy to the touch. These clays may be made up of meteor dust, volcanic ash, and severely decomposed organic material. Other sediments are authigenic, minerals that are crystallized out of the sea water itself. The most dramatic of these are manganese nodules.

When deep ocean deposits are discovered to be more than 30 percent composed of organisms, they are called oozes, pelagic oozes. Many oozes have brown clay, and they are soft but full of the remains of myriad microscopic organisms.

Around the borders of most of the continents, the terrigenous sediments are mostly silty clay. These sediments have few organisms because the material from land is introduced rapidly. The muds can be of different colors, depending on their land source; blue, due to the presence of some organic matter and altered iron compounds, black in basins without much water circulation, full of organic compounds and iron sulfide. Black muds usually smell of hydrogen sulfide. (The bottom of the Black Sea is like this.) Red mud comes from iron oxide introduced by rivers. Green muds are very much the same as blue ones. Some tropical muds are white, almost all calcium carbonate debris from reefs.

Some deep ocean deposits are full of sand and coarse silt. Many are believed to be due to turbidity currents that may have flowed as far as a thousand miles from land. Terrigenous sediments also include the matter that is ice-rafted on the bottoms of icebergs from glaciers and shards of glass and layers of ash from volcanoes.

Over the ages, these sediments have accumulated on the ocean floors in many different combinations. Some sediment layers are very thin (on the Mid-Atlantic Ridge almost nonexistent), while others are more than a mile thick. Recovering a whole undisturbed sequence of sediment layers, from the ocean floor to the original basement rock, is, of course, the new *Challenger's* objective. Analysis of the cores is done by all the means mentioned, but the richest, most subtle of the methods are those of invertebrate paleontology.

What pale and dry words they are. Invertebrate paleontology! Such a varied and fascinating world lies buried underneath that dreary term, a vast realm that many people do not even know exists. Invertebrate scientists may be the original models for the comic-strip picture of the fussy old scientist who spends his whole life bent over a microscope looking at the fossils of just a single species that lived during only one epoch of all geological time. To the uninformed, such paleontologists may seem the most remote and unworldly of all scientists, but actually these men and women live in tiny worlds of wonderful complexity, of awesome numbers of centuries, and what they know is crucial to understanding the history of the world. Moreover, many invertebrate paleontologists draw extremely good salaries from the men in oil who decide where the money has to go.

In ocean sediments, invertebrate paleontologists care most about protozoa, the simplest members of the animal kingdom. Protozoa are mostly single celled animals, but this cell may be distributed through many chambers in a complicated skeleton. They have no digestive organs and some, like the famous *Amoeba,* simply eat by flowing over a tiny plant cell and absorbing it. Usually protozoa reproduce by fission, the normal process of cell division that means growth in higher animals, but some are also capable of sexual reproduction. Protozoans live in all kinds of water, in moist earth, in the digestive tracts of most animals and even, as malaria, in the blood streams of human beings. Numerous larger animals depend on certain protozoa as their major food source in water environments. What paleontologists like about certain marine protozoan is that they grow shells out of the calcium in the sea. Protozoan shells, or tests, have rained on the ocean floor in such abundance that they have become important creators of rock.

One widely found protozoa is coccoliths, particles only 1 to 15 microns in diameter. They are also described as "animalcules one one-hundredth of an inch wide." A micron is one-millionth of a meter and a meter equals 39.37 inches. Coccoliths require a magnification of 400 times before they can be studied and so until recently their tiny, roundish skeletons were neglected in practical fossil work.

The "Summary of Coccolith Biostratigraphy" on Leg I, by David Bukry of Scripps and the U.S. Geological Survey and M. N. Bramlette of Scripps, shows that this fossil is gaining popularity.

"One great value of Coccoliths to deep sea research is the speed with which they can be used to date a sample. A simple smear slide, containing more than 100,000 coccolith specimens, can be prepared and examined under a microscope within a few minutes. Thus, aboard the ships it was possible to identify diagnostic species and determine the age of a core within five minutes of its recovery.

"An additional advantage of coccolith dating is based on their small size. Even the smallest amount of sediment can be dated to add to the geologic information about a drill site. For instance, at Hole 4A, which was drilled to recover relatively deep Cretaceous sediment, the first coring was not attempted until after 72 meters (236 feet) of subbottom penetration. The three cores then recovered were all Cretaceous in age. At the completion of this hole, however, when the drilling pipe was brought aboard the ship, a small amount of sediment adhered to the bit. This was found to contain a coccolith assemblage of the mid-Tertiary *Triquetrorhabdulus carinatus*. Even though this sediment was bypassed during the initial drilling, important additional geologic information was obtained when the bit accidentally struck the side of the hole on withdrawal."

Discocasters are minute bodies belonging to the same order as coccoliths and they may actually be plants. No discocasters of recent times are known for certain and so they remain on the borderline, but their disc and star-shaped bodies are useful for fossil work.

The luminaries of the protozoan world are the order known as foraminiferida. As "forams" they are the darlings of the oil industry. Many workers act as if the study of forams were the whole science of paleontology. Micropaleontologists who know the relations of various foraminiferal zones in documented oil-producing strata can predict, by recognizing a known strata in a new well, the distance down to a strata that is known to produce oil. Most forams are so small that they are not all ground up when the teeth on the drilling bit cuts into rocks and they come whole to the surface in cuttings and are easy to study.

Knowledge of forams increases at such a rate that an

illustrated catalog of them increases by 300 to 500 pages every year. The volumes now make up a row of books 20 feet long. Probably 30,000 species of forams are recognized today.

Not a great deal is known about the life cycles or diets of forams. It is difficult to keep them in captivity. Most of them live near the surface, but some have been found in the abyss at 13,500 feet.

Some forams are visible to the naked eye. They first occur in written history when Herodotus, in the fourth century B.C., mentioned their rounded shells in the stones of Egypt's pyramids. He thought they were fossilized lentils from the workers' soup. Many foram shells are washed up on beaches that have no surf. These mix in with sand grains of the same size and were recognized on the Lido beach of Venice by an eighteenth-century Italian, Gualtieri. A powerful microscope is needed to study most forams.

As the animals grow, they build successive chambers and the chambers are connected to each other by little holes. The living protoplasm usually resides in the last chamber. The animals got their name from the Latin for "little hole bearers."

Most fossil foram shells are calcareous, otherwise often called limestone or calcium carbonate. They are found in a tremendous variety of different shapes and the differences, naturally, determine the species. Some confusion is caused in naming species since shapes may change as the animal matures. Under a good microscope forams are quite beautiful, some pearly white, some chalky, and ornamented in almost every imaginable way.

Though all forams are important, some are more important than others and then they are known as index fossils. One such strain is called fusulinids, a group that

began back in the Devonian period, more than 300 millions years ago, and ranged through most of the world
and completely dominated the protozoan ranks for about
50 million years, more or less, at the end of the Paleozoic
era. Some strata of sedimentary rock on land are made
up almost entirely of fusulinids. The highly successful
fusulinids are now extinct.

Another superfamily of forams contains the globigerina,
probably the best known foraminifer of all. It was in
cold, globigerina ooze that the first *Challenger* cooled
its champagne when the voyage needed a toast in 1872.
According to Thomas Henry Huxley in his essay about
fossils, "On a Piece of Chalk," globigerina "looks like
a badly grown raspberry." White globigerina ooze looks
like a pudding.

Forams are so good as fossil indicators because they
are so numerous. Most of the evolutionary changes within
the species show no obvious function, no better adapta-
tion to their environment. In fact, the most complex
species always died out.

Forams are also supposed to be fine indicators because
they were so sensitive to climate changes. A difference
of one or two degrees would cause changes in the species.
This temperature sensitivity would make them coil to
the left in warm water and to the right in colder water
but exceptions to this rule have been found among fossil
forms in the North Atlantic. One virtue of forams is that
the species were so ubiquitous around the world they
may be used to synchronize the times between geologic
formations that are widely separated.

Individual forams may be seen in sediments as tiny
white specks. Micropaleontologists call them "bugs," as
they do the rest of their fossils. In some cores, fishbones
and very occasionally shark's teeth are found along with

forams. To prepare forams for the microscope they must first be rinsed through fantastically fine screens to get rid of all the clay and silt and baked in a special oven, slow at about 90°F., until dry enough so that the individual shells can be separated on the slide with a fine camel's hair.

Calcium has been widely used by nature as a building block but paleontologists find this choice of material something of a drawback. Under certain conditions it dissolves. Calcium carbonate is found in marble, limestone, coral, chalk, and shells. It is the fifth element in world abundance and very important in plants and the human body. But as the shell of dead protozoa sink through water that is undersaturated with calcium carbonate, the shell loses the chemical gradually to the environment around it. If the bottom is shallow, the shell will reach the bottom safely and be buried before it has been dissolved. If it has to sink in deep water, however, below what is known as the depth of compensation, it will cease to exist.

In such deep cases, paleontologists often look to a protozoan fossil known as radiolaria whose shell is based on silica. Ocean water dissolves silica very slowly. Seen through a fifty power microscope, radiolaria are some of the most beautiful constructions in all nature. They are either spherical or helmet-shaped, and very often they have long, delicate spines that radiate from the center. The variety of radiolaria ornamentation is so great that they are one of the most diverse groups of animals known. The first *Challenger* described 4,217 species.

Radiolarians are among the oldest fossils ever discovered. The earth of the island of Barbados, composed of radiolarians, long puzzled geologists who insisted that the ocean basins are permanent. The *Challenger*'s Leg

IV would investigate this subject. Fossil radiolaria are boiled hard in hydrogen peroxide, before study under magnification, to dispose of any calcareous material.

The time scale of today is based almost entirely on sediments deposited in near-continental environments and now exposed on land. These formations are often disturbed and discontinuous and correlations between continents are highly controversial. Good fossil sequences in true deep sea environments, found by the *Challenger*, may provide a reliable time scale that all scientists can agree upon.

8

"Now man can take his destiny in his own hands. Man once behaved in a very planktonic way. He voyaged downstream. He sailed with the current. Captain Cook found the Hawaiian Islands only two hundred years ago because they were not in a main stream. Now we sail upstream, horizontally and vertically. This requires profoundly more energy."

—MELVIN N. A. PETERSON,
Chief Scientist of the Deep Sea
Drilling Project

SEVERE FRUSTRATION and great success both marked the second episode in the *Glomar Challenger*'s scientific journey around the world. The ship had sailed from Hoboken, New Jersey, on October 1, 1968, after changing crews, receiving quantities of stores and new technical gear, and after a warm reception by the press, television, and members of the scientific community. The new Verrazano Bridge at the Narrows of New York Harbor had been built high enough so that the largest passenger liners and aircraft carriers had plenty of room to pass beneath, but the bridge engineers had never planned on a towering mobile derrick like the *Challenger*. The ship slipped under the Verrazano with only a few feet to spare and it seemed to sightseeing scientists riding in the crown of the derrick that they would certainly strike the under part of the bridge.

Leg II had Dr. Melvin N. A. Peterson as the senior of its two Co-Chief Scientists. Dr. Peterson had been Chief Scientist of the entire DSDP since the previous December and, along with many other responsibilities, was concerned with the selection of scientists who would go along on the cruises and on the very complicated question of deciding just where on all the millions of miles of ocean floor the ship would do its drilling. Choosing the perfect scientists and the ideal scientific sites required great diplomacy. Peterson had to see that the five allies (the participating scientific institutions) were all properly represented and that their regions of special concern and also their pride were properly taken care of. In addition, he had to make sure that scientists from other nonfederal institutions and scientists from government agencies interested in oceanography were also involved in the cruises. That was, after all, a national project and financed by the federal government. For scientific and diplomatic reasons, Peterson also tried to secure a good representation of scientists from foreign countries, particularly when the ship was working in waters near those countries.

Practically every man with knowledge in the field had his favorite part of the ocean where he wanted to drill and each man had very persuasive reasons for suggesting that his own site would be a very rewarding one. Probably all the suggestions had merit but sites had to be chosen not only in terms of scientific interest but also in terms of time (How long would it take to get there, did it fit in with other plans, and how long could one cruise last?); of weather (It was decided to keep the *Challenger,* as much as possible, in waters where the weather was traditionally good during the particular season); the ship's capabilities (The site had to be in deep sea but not much more than 20,000 feet and it could not drill through more than 2,500 feet of sediment or on bare rock without enough sediment to hold the drill assembly); politics (This became particularly important as the ship moved away from United States waters); and, finally, the simple fact that many sites met all the other requirements and seemed equally valuable. (Just to add problems, the site also had to be reasonably flat.) Any site might cost several hundred thousand dollars to drill and the decisions were as hard to make as Solomon's. Dr. Peterson had to be not only a scientist, but also a diplomat and a juggler.

Mel Peterson—rather tall, blond, and intense—was born in Northbrook, Illinois, and attended Northwestern University until he received his master's degree in geology in 1956. He did his military service in the U.S. Navy as an officer on destroyers and auxiliary ships for three years. He received his doctor's degree in geology from Harvard in 1960, and joined the staff at Scripps the same year. An oceanographer is a scientist, of course, but he is certainly better off if he is a sailor as well.

Dr. N. Terence Edgar, the Co-Chief Scientist on Leg II, has an accent that it would take Professor Henry Higgins to diagnose. Terry Edgar's mother was Scottish and he grew up in the English Midlands, where his father was a designer and manufacturer of aircraft. Terry, himself, is a licensed pilot. He was an undergraduate at New England's Middlebury College and married a girl from Tallahassee, Florida. He has done considerable geologizing in the Arctic regions. He did graduate work at Lamont, where he sailed on oceanographic cruises, and received his doctorate from Columbia on the undersea geology of the Caribbean. From Lamont, he went to Scripps and there he became Coordinating Staff

Geologist of the Project six months before Leg II began. When he goes ashore at a foreign port, his colleagues are surprised when he shows a British passport.

Although work on the *Glomar Challenger* prevents him from holding classes now, Terry Edgar is that rare kind of human—a born teacher. He sparks with interest at whatever question is asked or comment made, is off and running with every new idea, and yet seems to be looking over his shoulder at the same time to be sure that the person he is talking to is keeping up. He is extremely solicitous of those who work for him or study under him, and in his conversation he very frequently refers to the bright people around him. He looks as little like the musical comedy professor as possible, but more like someone in a leading role. Neither big nor small, he walks on the soles of his feet like a boxer. He seems to relish helping to solve the constant mechanical and technical problems that arise on a drilling ship at sea. Research scientists on location are called upon for much more than their book learning. When crucial equipment breaks down, they feel that they are wasting time and so the impatient scientists soon learn to pitch in and correct matters. Of course, most essential scientific equipment was planned by the researchers themselves because only they know what is needed. Today, young scientists like Dr. Edgar have to know a great deal about technology and the capacities of electronic equipment and computers if they are to pursue their own specialties.

In connection with interviews, Terry Edgar has spoken of his distress when a reporter quoted him as saying that the rising material at the ridge *pushed* the ocean floor away. "I don't know how he got that out of what I told him," said Edgar. Spreading is not to be understood as a pushing motion but visualized as similar to a conveyor belt that is carried along by a force underneath.

Leg II also had three other men scientists from Scripps, one from Miami, one from an engineering firm, and two ladies, Dr. Catherine Nigrini from Canada and Dr. Maria Cita from the University of Milan, in Italy. Mrs. Nigrini is a specialist in radiolarian paleontology, and Mrs. Cita in foraminifera.

Catherine Nigrini is young, pretty, and married. She enjoyed Leg II so much that she asked to join the ship again when it sailed in the Indian Ocean. Handsome Maria Cita of Milan has two teen-age boys and when Americans in the field were asked to recommend a European woman paleontologist for the cruise the reply was simple: "You should ask Maria Cita." She was written to and replied enthusiastically. During the seven weeks at sea, the two women, who had never met, pitched in to do hard, scrubby work in the labs when the going got tough, and both had a wonderful time. When Dr. Cita returned home she wrote happily about the *Challenger* both for the European press and for technical journals, and she rejoined the ship when it sailed from Lisbon into the Mediterranean on Leg XIII.

Just before the ship began Leg II, the Atlantic Advisory Board changed the plans. A site north and west of Bermuda, that should have been covered on Leg I (but could not be for lack of time) was to be included as the first stop on Leg II. Within Project circles, this became, in time, the infamous "Site 8," a place on the sea floor about which the farmer might have said, "You can't get there from here."

To recapitulate, the basic purpose of the *Challenger* drilling is to find out the age of the oceans and explain how they were developed. In the old tradition of geology, the ocean basins were believed to have been where they are now since the time of creation. When the idea of continental drift came along in 1912, an idea which meant that the ocean basins had certainly once been different, it created quite a stir. According to Dr. Robert Dietz, the man who published the term "sea floor spreading," one University of Chicago professor regarded drift with such hate that it is said he became incoherent for several days if the concept was even mentioned in class. A prominent Stanford professor published a paper entitled "Continental Drift: A Fairy Tale," hardly a suitable title for a scientific dissertation. Nevertheless, drift and changing ocean basins have made their way slowly into the scientific world.

If the oceans have never changed since the original creation, then the bottom sediments should contain a more or less continuous record of the whole geological history of the world. But they do not. Therefore many scientists now are willing to believe that on a scale of geological time the earth's surface is very fluid, that continents do move about in relation to each other, and that the ocean floor is, in terms of hundreds of millions of years, very young indeed. Deep drilling was, first of all, to verify this new thought.

In addition, deep drilling was to help correlate and standardize the time sequences of a number of related sciences. For years micropaleontologists had been establishing time scales for their tiny fossils based on their evolutionary changes. These fossils, of course, are found in enormous abundance in the ocean sediments. Geophysicists had also been dating sediments in terms of the age of the magnetic reversals to be found in their mineral content. Dating was also being done by means of chemistry and mineralogy. Long cores from the deep sea floor representing extremely long periods of time could be taken by the *Challenger,* and the material could be studied by specialists in all four fields so that one could know, for instance, that such and such a fossil was associated with a particular time on the magnetic reversal scale and that at this particular point in time you could expect certain chemical and mineral conditions. A complete correlation of these various factors for different parts of the world would provide a very detailed history of the world's past. Although such a correlation will take years to put together, long sequences of microfossils at one location, which would show the evolution there, could be related to the fossil evolution story at other locations comparatively rapidly.

Beyond this kind of work on sediment, it was hoped that drilling would provide information on where various inorganic sediments came from, how they arrived at their point of deposit, and how they formed new combinations such as chert and manganese nodules.

The Atlantic Panel thus proposed thirty sites that might yield the ocean's oldest sediments, test sea floor spreading and continental drift, or produce a complete sedimentary column. No one site was likely to be ideal for all three objectives.

To choose the thirty sites the panel had numerous reflection profiler records and data from piston cores made during a great number of oceanographic cruises. The information most often used came from work done by ships from Lamont and Woods Hole.

In the Atlantic it was known that the thickest and so, presumably, the oldest sediments lie near the continental margins, but these old sediments were far too deep for the *Challenger* to drill. Therefore, sites were looked for where it was known that older sediments were close or exposed on the ocean floor. Such a site, east of the Bahamas, had been drilled on Leg I because short piston cores had previously found sediments 120 million years old near the floor surface at that location.

Earlier surveys had shown that sediment in the Atlantic gets thinner, quite symmetrically, as you approach the center. There is almost no sediment on the peaks of the Mid-Atlantic Ridge which occupy the middle third of the ocean. The Ridge, as has been said, is part of the longest mountain chain in the world, extending under the sea for about 40,000 miles. In the Atlantic, and only in the Atlantic, it is equally distant from the continents on either side.

"In addition to these symmetrical characteristics," Peterson and Edgar said in a paper they wrote together in 1969, "a pattern of magnetic anomalies, that is, variations in the earth's magnetic field attributed to the nature of the crustal rocks, parallels the linearity of the ridge and forms a symmetrical pattern about the ridge crest. These characteristics have led geologists to believe that the ocean floor is spreading away from the central ridge crest and that the earth generates new ocean crust along the ridge axis.

"If this is the case, the youngest and thinnest sediment should be at the ridge crest. Nearer the continents the sediments should be older and thicker. In general, you can see the distributional pattern in sediments from the profiler records. But no one could establish the age of the oldest sediments at any given location between the ridge crest and the continents from profiler records.

"The Atlantic Panel therefore recommended drilling a series of holes across the ocean to determine whether the oldest sediments overlying the 'basement' or igneous rocks, are indeed younger nearer the crest of the ridge."

Peterson and Edgar put the word "basement" in quo-

tation marks as a scientific precaution. They were refer-
ring to the deepest reflection you can get from a profile.
Presumably this is the earth's volcanic crust and no sediment exists beneath it but how can one ever be sure that
lavas have not simply rolled over old sediments, present-
ing such a solid obstacle to sound waves that what seems
to be the true bottom—formed before any sedimentation
at all—may only be an apparent bottom and that many
millions of years lie concealed beneath it? There will
always be the sneaking possibility that this deepest reflector, this basement you have even drilled into, may not
be the primordial crust itself, and so the most complete
"proof" of the age of the ocean bottom can never be quite
absolute.

It might also be noted here (but not for the last time)
that the magnetic anomalies, the magnetometer readings
that show the magnetic poles reversing, and which are
so important to the proof of continental drift, are only
strong for about 75 miles on either side of the Mid-
Atlantic Ridge. (This represents five reversals.) Farther
out than 75 miles, the amplitude of the magnetic signals
weakens and near the continental margins it is quite hard
to detect. Those who still resist the drift idea cling
strongly to this fact. Lamont geophysicists who first
pointed out the magnetic anomalies and their relation to
drift argue that the low amplitudes in the western
Atlantic, from the Grand Banks to the West Indies, exist
because 200 million years ago the alignment of the area
was so different that it then represented the magnetic
Equator and so weak magnetism was put into the molten
rocks. Or, if that explanation is not accepted, they propose the fact that, at the time of rifting, the continental
margins were close together, sedimentation was rapid,
and the magnetized rock particles were thoroughly

buried to a depth below which signals are muffled.

Objections to any proof of drift were anticipated by the Atlantic panel. "In order to examine the concept of sea-floor spreading, where magnetic anomalies are not easily correlated, it was suggested that an age-spatial relationship be established by paleontologically dating the sediment immediately overlying the acoustical basement at a known distance from the ridge axis. *A decrease in the age of the sediments toward the ridge crest would support the concept of sea-floor spreading.*" [Author's italics]

If the doubters balk at the proof of drift by magnetic changes, let it be proved by showing that the most deeply buried tiny fossils become increasingly younger as you approach the center of the ocean, the point of origin for the spreading.

It seemed to the panel that a site north and west of Bermuda might produce some of these oldest sediments in the Atlantic. Drilling might also discover some of the oldest available rock, and cores of the sedimentary column might show how the Gulf Stream, whose constitution has changed over the ages, has influenced the marine ecology of the area.

And so the notorious Site 8 was born.

The *Challenger* arrived on the location 35° 21' N, 67° 31' W (which had already been scouted for by Lamont's little ship, *Vema*), at 5:42 P.M. on October 2. Almost at once the positioning system began to malfunction and the ship drifted off its station even before the acoustic beacon could be lowered. At 11:58 P.M., final positioning was achieved. Horizons A and Beta, the reflectors that had been sampled on Leg I, were found again in this location. The drill string was assembled with a diamond bit and the sea floor was entered at 11:00 P.M. on the following night. First, 521 feet of soft sediment were drilled and then a 30-foot core was cut, but the inner core barrel was brought up empty. Another core was attempted immediately and it contained 3 feet of gray-green radiolarian mud of upper Middle Miocene age, somewhere around 30 million years old. Drilling was begun again and at about 800 feet below the sea floor, the drill went through two hard thin layers. At 817 feet, a core was cut again. The men on the surface could only guess what went wrong somewhere down the drill string but, whatever it was, the sand line that retrieves the core broke and all that could be done was retrieve the whole drilling assembly. The bottom of the barrel had 12½ feet of sediment in it. The radiolaria in it were from the Eocene epoch, about 60 million years ago.

The scientists calculated from seismic reflection records that they had reached a depth about 100 feet above the top reflector. They decided to try again to sample this reflector. The drill string was reassembled, and drilling proceeded to a depth of 913 feet below the sea floor where resistant layers were encountered. Chert was suspected. It was thought that perhaps continuous coring, rather than drilling, might be a better way of getting through the chert. On October 7, beginning at 1:00 A.M., three cores were recovered, with a total yield of about 12 feet of sediment, much of it chert. With the fourth coring attempt, drilling suddenly seemed to go rapidly and it looked as if the chert had been penetrated. At 5:30 P.M., this fourth core was retrieved, but the core-catcher was missing and no core was recovered. The core-catcher had become detached from the inner core barrel, and remained at the bottom of the drill string, and thus the core could slide out. Since there was probably chert above the core-catcher, no attempts were made to save it.

Furthermore, the lower part of the bottom hole assembly and the diamond bit were lost. At 9:30 P.M. on October 8, the *Challenger* was underway to Bermuda.

During the six days of agony on Site 8, a total of 29 feet of core was recovered. The water was always very rough and the wind sometimes blew at 45 knots. During much of the time, one bow thruster failed to work and, in the heavy seas and wind, the other bow thruster had to do the work of two. Edgar recalls many times hearing the overworked machinery scream in agony. The chert, of course, was an unseen but implacable enemy. In addition to everything else lost in the bottom assembly, a quantity of valuable electronic tools were down at the bottom at the time and they went too. Site 8 was given up, finally, when the bow thruster bearing absolutely disintegrated. Edgar, who had been up for most of the six days, lost five pounds. From Bermuda, the *Challenger* had to go to Norfolk for five days in drydock so that the bow thruster could be repaired. With one possible exception, Site 8 was the worst that *Challenger* scientists have so far experienced.

When once more ready for sea, after repairs in Norfolk, the *Challenger* sailed again to find out the history of the North Atlantic Ocean. Site 9 was on the flank of the Bermuda Rise, several hundred miles east of the island itself. The bottom in this region consists of low lines of ridges that run in a northwest-southwest direction, there are scattered seamounts with peaks 12,000 feet below sea level, but the sea floor is otherwise flat. The rise itself is dissected by deep valleys with steep flanks. The Sohm Abyssal Plain lies between the site and the beginnings of the Mid-Atlantic Ridge. The water at the position was about 16,000 feet deep. This eerie seascape had almost no seismic reflectors, no Horizon A or

Beta, and was chosen to examine a good sedimentary column for study in relation to sea-floor spreading. The basement lay below something between 2,600 and 2,800 feet of sediment. The top few hundred feet of sediment reflected acoustic signals and then it was transparent, except for one weak reflector, all the way to the bottom.

The program was to core continuously until the transparent sediment was reached, then core at intervals through this long section, take a sample of the weak reflector, core continuously between the weak reflector and the basement, and then attempt to core the basement rock itself. This last step had never been done before.

The first core was recovered at 5:00 A.M. on October 22, but was found to be empty except for traces of sediment in the catcher and liner. The core seemed to have been cut but the sediment was apparently too soft to be retained by the catcher. The second, third, and fourth cores all produced the same discouraging, negative results. Smear slides of traces of sediments showed that the fossils were not much more than a million years old. The fifth core did produce 25 feet of sediment but it was so highly disturbed that no time sequence could be made of it. The sixth core was again lost through the catcher.

The men in charge decided to stop continuous coring, the center of the drilling bit (tungsten, this time) was replaced, and drilling progressed down to 630 feet below the ocean floor where the sediments seemed to be a bit more solid. Cores seven through twelve produced only small amounts of Miocene-Pliocene clay, 10 to 30 million years old. (It is wonderful to consider how geologists can reach so far back in time and then casually dismiss the result as unimportant.) At 1,613 feet beneath the sea, the power sub failed and the drill string became stuck in the hole because the pipe could not be rotated. After a struggle, the pipe was freed and the hole abandoned.

A new hole was then begun very nearby. The goal was now reduced to getting the oldest sediment and reaching the basement. The first core was not taken until a depth of 2,226 feet into the sediment had been reached (half a mile). The core this time was Eocene, chert, and radiolarian ooze 60 million years old. This core was very difficult to get out because the plastic liner had been deformed, probably due to overheating of the bit. Drilling for the second core in this hole went several hundred feet farther down before another sample was taken—about one foot of hard clay with some coccoliths, radiolaria, and foraminifera, little animals from the upper Cretaceous. Again the plastic liner was deformed and discolored cloudy white, probably due to overheating. Then a third core was cut, seven feet of clay, and the power sub failed again. The center bit was replaced and drilling progressed by means of the rotary table and kelly on the ship's drilling platform. The seas were fairly high and at one point the kelly jumped out of the table and some of the gear was bent. Because the liners were being deformed, the rest of the cores were taken without them and the sediments brought up by hydraulic pump.

At 2,740 feet the basement was reached and 200 grams of igneous volcanic rock were recovered from the lower portion of the core, which was only about a foot long.

At this point the *Challenger* had drilled more than 200 feet farther down than it was supposed to or designed for. Risk was high that the drill string would be damaged or even lost, or that the hole might collapse, so the drill string was brought back to 2,500 feet before an attempt was made either to recover the core or log the hole. This drilling was accomplished at the end of the longest drill string ever suspended from a floating platform, 19,075 feet, and the sample from the bottom was reached at the deepest penetration ever achieved into the floor of the deep ocean. The last sediment was brick red, a symptom of iron. The Chief Scientist found this iron oxide to be of considerable significance.

Moving eastward across the Atlantic, Site 10 for the *Challenger* was chosen on the lower western flank of the Ridge. Before mechanical problems cut short the time and forced a revision in plans, it had been the westernmost of a series of sites proposed to cut across the axis of the ridge. The major purpose was to check the concept of spreading. During the drilling, some cores were lost due to the catcher; there was some mechanical trouble, too, but the basement was reached and, during 122 hours of work, 251 feet of core were recovered. The greatest sediment age here, about 500 miles west of the ridge, was Upper Cretaceous, or perhaps 85 million years old. Recovery of only 44 percent of the core drilled was not considered very good.

The basement was reached again on Site 11 and the sediment above it was no more than Middle Miocene, 18 million years old. The site was on the western flank of the ridge, only 100 miles from the crest itself, so the most important point was proved. The deepest sediments do certainly become younger as you approach the center point where spreading is presumed to occur.

Except for the Miocene fossils, Site 11 is not a fond memory for the scientists on the *Challenger*. The sea floor was penetrated to a depth of 932 feet and eight cores were taken, but their total length (each core barrel is 30 feet long) was only 22 feet. This is an average recovery of 13 percent. Only the fourth, seventh, and last cores contained anything for the specialists to look at. Something was very wrong with the core-catcher.

As the *Challenger* steamed toward its last position on Leg II, Mel Peterson thought very hard about the problem of core recovery. He felt considerable distress over the enormous loss of valuable sediments, the research material that was the reason for being far out at sea and weeks away from home. Not long before the ship arrived at Site 12 just north and west of the Cape Verde Islands off Africa, he conceived of a plastic sleeve that could be attached to the core-catcher that might save more of the soft sedimentary material. With the help of a technician, he had it built before they settled in on the chosen location. Fourteen days had been scheduled for Site 12, but only six days were left by the time they arrived. Here, close to Africa, the sediments were more than 2,000 feet thick and the object was to obtain material for the paleontologists to study. The evolutionary sequences of foraminifera have been established for the Caribbean, as well as for western and southern Europe, and at Site 12 it was hoped to provide a link between these two sequences and to show correlations of fossils forams discovered at other sites across the Atlantic. It was hoped, too, to find fossils as old as Jurassic, 150 million years ago, the age of the sediment taken in the Bahamas during Leg I. Here Mel Peterson's new invention was used and the core recovery was a great deal better. Over 700 feet of cores were taken at the site. Some of the sediments were so soupy that the cores had to be frozen in order to preserve them. Nothing as old as Jurassic was found.

Chert was found at this site, too, and its presence in so many places in the Atlantic led Peterson and Edgar to believe that it was caused during the Eocene time, about 65 million years ago, when the Isthmus of Panama was under water and Pacific water flowed in and mixed with Atlantic water, changing the circulation and chemistry so that this hard substance could form.

The major statement for Leg II of the *Challenger*'s career, however, was that they "have found important evidence strongly substantiating this elegant and bold hypothesis called 'sea-floor spreading.'"

Terry Edgar's personal statement about the voyage was that "We found that all our main ideas about the bottom were true but few of the details were."

9

When Michael Faraday, the English scientist who invented the electric dynamo, showed the machine to an acquaintance he was asked, "Well, of what use is it?"

Faraday answered, "Well, of what use is a new-born baby?"

IT WAS the picture of South America and Africa on a map, the way they seemed to fit so neatly together, that first gave people the idea that the two continents could once have been joined together and then separated. Therefore, the scientists decided that the *Challenger* should test the drift idea in that part of the ocean as soon as possible. From Dakar, the ship was in the right position to make a sweep across the South Atlantic, drilling holes, with the attractive port of Rio de Janeiro waiting for them when work was done. First the *Challenger* spent a week in Africa taking on supplies, changing to the alternate crew, and receiving its new team of scientists.

The leaders were now Dr. Arthur E. Maxwell and Dr. Richard P. von Herzen from Woods Hole, in Massachusetts, the oldest oceanographic institution on the East Coast of the United States. Along with these scientists were geologists from Zurich (Switzerland), Hawaii, and Scripps, as well as paleontologists from Lamont, Scripps, and Princeton. No ladies took part on this cruise.

Compared to the tribulations of the previous two

months, Leg III was technologically a great success and, in terms of science, it could be said at its close that now "Almost unassailable evidence confirms sea-floor spreading."

The first site, Site 13, was on the Sierra Leone Rise, in the eastern Atlantic, off Africa, where it was hoped that the thick sediments would provide many strata, or levels, of geological and fossil material that could be linked together with similar levels in the western Atlantic, a region much more thoroughly studied. It was also expected that Site 13 would produce oozes of radiolaria from Eocene times, such as had been found elsewhere in the Atlantic. As has been said, Pacific water is much richer in silica and the radiolaria that use silica as building blocks. An increase in Atlantic radiolaria might mean that Pacific water had been circulating in the other ocean. A wide variety of sediment types was encountered and the sequences were not much different from those in holes in the western part of the North Atlantic. The supposed change in circulation may indeed be proven, since the predicted Eocene oozes were found. These oozes, full of radiolaria, strongly suggest that a current near the Equator ran between the Pacific and Atlantic about 60 million years ago.

The site was one that had been surveyed by the *Vema*, and the *Challenger* determined it easily by satellite navigation. The weather was ideal, all the positioning worked well, and drilling and coring went on for six days until chert destroyed first a diamond bit and then a tungsten bit.

As Maxwell said in a speech about Leg III, "To give you some indication of the problem of drilling in the deep sea, it is analogous at a reduced scale to the problem of lowering a long drinking straw into a coke bottle from a helicopter, bouncing in the wind, at the height of the Empire State Building. Even if you were lucky to hit the bottle on the first lowering, it would be highly improbable to go back into the same bottle again. Therefore, the limitation of any one hole is when the bit wears out or some other failure causes you to cease drilling at that point."

From Site 13, the *Challenger* moved onto the most basic objective of the leg—to straddle the Mid-Atlantic Ridge with a series of holes dug on either side. Here every operation went smoothly and all hands worked very hard, even taking seven cores on Christmas Day.

Maxwell's explanation of the choice of drilling sites is one of the best in making clear what may seem like a somewhat abstract idea.

"Hot molten material from the earth's interior rises to the surface in the middle of the ocean forming a feature known as the Mid-Atlantic Ridge. Once this molten material hardens, it is replaced by new lava from below, at the axis of the ridge, and the solidified rock moves slowly away from the axis in both directions—as if the blocks were on two conveyor belts moving slowly in opposite directions.

"Thus, the Atlantic Ocean appears to be growing at the expense of the Pacific because South America is overriding the Pacific. This hypothesis of sea-floor spreading was first suggested by the late Professor Harry Hess of Princeton in 1960. At first, scientists, as they are naturally inclined to do, did not accept this concept. However, as often happens, measurements made in another connection shed new light on the hypothesis.

"At about the same time some geologists on the West Coast noted that lava from volcanoes became magnetized weakly in the direction of the earth's magnetic field when it solidified. Further, they noted that older lavas were magnetized in opposite directions. Additional measurements of the magnetism of lava, dating back to five million years in age, showed that the earth had experienced several reversals of its magnetic field during this time—the reversals lasting tens to hundreds of thousands of years.

"It took two bright students at the University of Cambridge, England—Fred Vine and Drummond Matthews —to put two and two together, and in 1963 they suggested that if sea-floor spreading were taking place at the Mid-Atlantic Ridge, the lava must have been cooling over long periods of time. Therefore, it should contain a record of the direction of the earth's magnetic field at the time it cooled. In effect, the Mid-Atlantic Ridge should act like a giant slow moving tape recorder that shows alternately positive and negative magnetism. An examination of magnetic profiles made across the ridge at many places showed this indeed was the case. Further, they found these magnetic anomalies, as they are now referred to, extend great distances away from the ridge axis. By correlating the positive and negative peaks of these results at sea with those measured from volcanic lavas, scientists were able to date the ocean floor back to five million years. Further, since the peaks continued beyond this time, by assuming the sea floor had been spreading at a constant rate, they were able to suggest the sea floor had been spreading for about seventy million years."

The area with the most well-defined magnetic anomalies lies in the South Atlantic. It was planned to drill holes in several of the magnetic anomalies at varying distances from the (ridge) axis where one should be able to predict the age of the basement rock. If the drilling confirmed the age of the rocks, or of the sediment immediately above, the case for sea-floor spreading would be considerably strengthened.

The *Challenger's* drilling back and forth across the ridge in the South Atlantic went so well that it was possible to drill two more sites, in the time before the ship was due in Rio, than had been originally planned. All of the sites were near the 30th parallel of latitude south of the Equator. The first lay about 400 miles west of the center of the ridge itself. The *Vema* had previously surveyed the location, a place where the water was 13,255 feet deep. In this part of the ocean, the magnetic anomaly, where a reversal had taken place, was somewhere between 6 and 9 miles wide. Here at Site 14 a high wind made it difficult to keep station automatically, a few mechanical problems occurred in the drilling apparatus, but coring went on almost continuously for two days and recovery of the sediment was 87 percent successful. After drilling in the basalt for 5 feet, penetration was reduced to zero during the succeeding two hours. The core was pulled and the hole abandoned at this point. The age of the sediment of the first site was at least 40 million years, which almost exactly agreed with the age predicted in the theory of sea-floor spreading. (An unexpected layer of white chalk was cored at a level above the basement and this gave the geologists cause for wonder.)

The hole had been drilled a bit west of Magnetic Anomaly 13 which, it had been suggested by Dr. J. B. Heirtzler of Lamont in 1968, should have an age of 38 million years. The sediments just over the basalt actually drilled contained a species of foraminifera that became extinct, according to Dr. William A. Berggren of Woods Hole, 39 million years ago. Thus, by bits and pieces, does knowledge often build sturdy structures. Not always, however, do two facts fit so neatly as the magnetism and paleontology of the *Challenger's* Site 14.

Site 15, almost 200 miles closer to the axis of the ridge, had a positive magnetic anomaly that, previous to coring, was said to be 21 million years old. While working at this position on Christmas Day, the wind went up to 30 knots and positioning was not easy, but by 6:00 P.M., the ship was back on automatic positioning and no other difficulties arose at the spot. This time 463 feet of cores were recovered; for the first time an almost 100 percent success. Nothing was lost, but only 1 foot of basalt could be retrieved, even after a 50-minute attempt. The deepest sediments at the next site showed an approximate age of 24 million years instead of the predicted 21 million years. This was not quite so neat a correlation with magnetism as at the previous hole but still quite remarkable.

Still moving in on the crest of the ridge, the ship next took a position 167 miles to the west of it after five satellite navigation fixes. This was right over a small hill, 11,325 feet below, and the drill, biting into it, found basalt 390 feet under the peak. Core recovery was again perfect, the scientists giving credit to Peterson's plastic sleeve core-catcher for saving all of the fairly soupy sediments. Nearer the Mid-Atlantic Ridge on the next drilling, the age should have been 9 million years but the fossils, Late Miocene, were more like 11 million years old.

On December 27 and 28, 1968, this part of the world ocean had swells that ran from 10 to 16 feet and a current of 1.5 knots. The ship sometimes rolled as much as 10 degrees and this tended to slow down drilling. The drilling speed varied considerably, probably not because the sediment changed from tough to soft but because the driller was being distracted by the motions of the ship. When the bit came up at last it was worn excessively and parts of it broken, most likely because it had bounced up and down on the basalt. Only one little chip of this rock was retrieved.

Operations having gone so well on Leg III thus far, Dr. Maxwell and Dr. von Herzen decided to take the *Challenger* across the ridge crest to the east side and see what symmetry they might prove in ocean-floor spreading. Their first extra, unplanned site was selected on what they believed to be Magnetic Anomaly 13, the eastern counterpart of the anomaly they first drilled on the west side. Site 17 had not been surveyed in advance and this was a bit troublesome.

The Chief Scientists' extra site had a steeply sloping basement. Drilling went along nicely beneath the sea floor for 304 feet, and then a layer was hit that stalled the pipe. Coring had gone well but a sample had not been taken of the sediment right above the point where the drilling was stalled so the pipe was pulled out and a new hole started, almost directly over the same spot. Here, however, the hard layer happened not at 304 feet but at 336 feet. This hard layer could not be drilled but fragments of basalt were brought up.

Consequently, the ship was moved less than a thousand feet away and the new limit on drilling came at 407 feet. This third hole was in somewhat deeper water. It was decided that probably only the third hole had reached the true basement, the other two holes reaching only intermediate hard layers, and the scientists concluded that another ship should make a detailed geophysical survey of the area to find out. The bottom fossils were dated as Late Oligocene and the basalt estimated as 32 million years, assuming that the ocean floor here has a constant spreading rate of 2 centimeters a year. The white chalk that surprised the researchers at Site 14 appeared here at a similar layer of the same age but the bottom layer here was 7 million years younger.

Moving in toward the crest of the ridge, from the east side, the *Challenger* men chose another unsurveyed region for their other extra site. The day after New Year's they began to put out about 13,000 feet of drill pipe, and core with almost perfect recovery until the diamond bit was chewed to pieces on the basalt. The region had never been studied to find a magnetic age for the basement, but computations gave a suggested age of 25 million years and the fossils recovered indicated 26 million years. As in other cores of this series, the sediments were studied to find out how ocean temperatures in this region had changed in the past. Some layers of sediment produced tropical species, other layers showed species of temperate waters. The details of these changes may take the infinitely patient paleontologists years to work out.

Scientists are looking for facts rather than high drama so the men on the *Challenger,* as they crossed over the ridge again toward South America to carry out the rest of planned Leg III, were quite satisfied with their very successful and quite uneventful detour. The next choice of the Advisory Panel had been 600 miles west of the ridge, on Anomaly 21. (If not already obvious, the anomalies were numbered outward from their point of origin, as theorized in the sea-floor spreading idea.)

Site 19 did just about what it was supposed to toward proving the great theory. Water depth being almost three miles, the drill pipe was started down at 9:15 A.M., shortly after the acoustic beacon for positioning had been dropped, and it reached bottom at about 6:00 P.M. The seismic profiler had been giving a dubious measurement of the sediment thickness, so two cores were first taken that, when analyzed by the paleontologists on board, gave an estimate of the rate of sedimentation. This knowledge, along with the expected age of the sediments at the bottom, allowed the scientists to predict arithmetically how thick the sediments would be. They estimated it would be 492 feet below the sea floor, which turned out to be correct within 5 percent.

After this little stunt, which probably no one ever taught them in college, drilling proceeded without much trouble until basalt stopped the drill at 477 feet. Sediments gave an age of 49 million years. An interpretation of magnetic anomalies had said 53 million years. There is not much magic in being able to produce such results, far away in the South Atlantic. It has only taken a lifetime of work by each of several hundred scientists of various disciplines.

The last drilling which was designed to prove that the South Atlantic has been spreading, separating South America from Africa, also showed that the men on Leg III had had extreme good luck on their cruise so far. Approximately 800 miles west of the ridge crest, over Anomaly 30, hypothesized on the time scale as having an age of 72 million years, Site 20 indicated that all problems in the technology of deep-sea drilling were not yet solved.

On location, first it was discovered that the acoustical beacon was not functioning properly at all and it became a critical problem whether there would be enough spare beacons to finish the leg. Then the winds blew from the NNW at speeds sometimes exceeding 30 knots, 1 knot currents flowed from the north, and there were long period swells from the south. Buffeted from three directions, position keeping became very difficult, the ship rolling and pitching while the signal from the beacon was very erratic. In spite of all this, an average of all satellite fixes gave an agreement within 98 feet, so they were confident of their navigation and exact location.

On bringing up the first core in the face of all these distractions, the bottom of the core barrel was found to have a piece of basalt lodged in it, consequently no sediment had been collected. Two more attempts to core showed that the drill string apparently had pieces of basalt stuck in it, pieces left over from the last site. The whole string had to be brought aboard and the debris removed. This task required 24 hours.

Then coring was begun again, in more than 15,000 feet of water, but, after 20 feet of soft sediments, the core jammed with pebbles of manganese. The decision was to move to a location with better bottom conditions. At the first hole here, basalt was found at a depth of only 197 feet. A second hole was dug in the same place to pick up the sediments missed on the first attempt. Now, while retrieving the second core, the drill string broke, leaving a core barrel, bit, bumper sub, and two drill collars in the hole. The whole drill string had to be brought back aboard, of course, and this took another 24 hours. The break was blamed on metal fatigue brought on while trying to core in basalt.

Finally, one good series of cores was produced without further difficulty, this time with a recovery of nearly 100 percent. The oldest sediments were dated paleontologically at approximately 67 million years and, since the anomaly would allow 70 million years, the agreement was considered quite good.

One "proof" that had been used against the idea of sea-floor spreading, at least in the South Atlantic, was that the *thickness* of the sediments was much the same all across the ocean. Leg III of the *Challenger* odyssey found this to be quite true. The argument was that, if the sea-floor were spreading and sediments were being deposited all the time, the oldest ocean floor, farther away from the axis, should have the most sediment. Since this was not true, what did the "spreaders" have to say for themselves?

An analysis of Leg III cores taken on both sides of the ridge gives an explanation. While all are more or less the same thickness, those far from the center have very few sediments from recent times. Those in the center near the crest have a great deal of recent sediment. For some reason, sediments are not being deposited or they are being eroded far out on the ridge flanks. Again, for some reason, the ocean bottom receives most of its sediments while the bottom is in the neighborhood of the crest. Why this should be is a new oceanographic mystery, but a consistent sediment thickness in the South Atlantic is shown to be no argument against spreading there, according to the Leg III scientists. Of course, in the North Atlantic, sediments are thinner at the center ridge and much thicker toward the continents and this difference has been used as an argument in favor of spreading. It all poses a problem but as far as Leg III is concerned, the comparative youthfulness of the deepest sediments near the crest and their increasing age as one moves away from it shows that Africa and South America are spreading apart.

The remaining sites planned for the voyage did not have quite the cosmic implications of the holes drilled athwart the ridge. As objectives, Sites 21 and 22 were primarily technical and of greatest interest to specialists. They were located on the Rio Grande Rise, a region off the southern coast of Brazil with relatively shallow water depths around 7,000 feet. One purpose was to recover the oldest sediments in this part of the South Atlantic and the other major objective was to examine the layers of sediment on the rise itself, partly in order to understand its ancient history.

Drilling was uneventful but some technical problems on the second site delayed matters so that cores that would have revealed the boundary between the Cretaceous and Tertiary—two great periods in geological time —could not be drilled for because time had run out. The *Challenger* was due in Rio de Janeiro.

The official report states that "valuable information concerning sedimentary processes has been obtained, and the cores provide the materials for research to further our understanding of oceanic sedimentation and diagenesis."

As a view of how the geology of the ocean is gradually being revealed by deep-sea drilling, the discovery of a high percentage of foraminifera leads to an interesting conclusion. As has been said, the shells of forams dissolve when they sink below the depth of carbonate compensation. This depth varies, but to the layman it seems like pretty deep water. In the 7,000 feet of water on the rise, the foram shells survived and at Site 21 could be dated back to Late Cretaceous, 70 million or so years ago. The drillers on this site could not reach down to any older fossils because the bit was stopped by coquina or limestone. This may indicate that the rise was even shallower in the past, possibly even being a beach at this point. Since the basement could not be reached, the question remains whether the rise is a fragment of continental crust left behind as South America drifted westward. Or it might be an oceanic basalt that has risen above the surrounding region as a group of seamounts or guyots, high enough for foram shells to survive.

An analysis of the calcium carbonate in all sediments taken on the leg suggests how this factor leads to geological understanding. Sediments on the east side of the ridge crest were 10 to 12 percent higher in calcium carbonate than those on the west. At present, the compensation depth on the east side is a bit more than 16,400 feet, while on the west it is 13,120 feet. This is because the cold bottom water that sweeps up from the Antarctic, on the African side of the ocean, is low in calcium carbonate and thus can absorb more of it. This cold water does not cross over the ridge crest or the Walvis Ridge off South Africa. East of the ridge, the water is relatively warm and sluggish and does not become undersaturated with respect to calcite until much greater depths are reached. Ocean circulation obviously has a great effect on geological formations in the sea, just as weather does on land formations.

As they studied the cores taken from the numerous holes they had drilled across the South Atlantic, the scientists aboard the *Challenger* found that the stratigraphy of sediment layers in the cores from one hole often resembled quite closely the stratigraphy in a core taken many hundreds of miles away. This was not the least bit surprising to them; the same thing occurs in land geology all the time. Stratifications of rocks of different kinds can often be seen on the sides of hills or mountains that have been cut away by natural means and in cuts made by man for roads and such. A classic example of stratification is the Grand Canyon where layer rests upon layer in cliffs a mile thick and many hundreds of millions of years of geological events are exposed. Such stratifications are known as formations. Certain ones have become classics. They are usually given the name of some nearby, fairly well-known geographic feature. A trained geologist knows the important formations and when such-and-such a formation is mentioned, he will remember its particular sequence of rocks. One such famous feature is the Green River Formation that can be seen exposed in Utah, Colorado, and Wyoming. Formations, or sequences, like this often recur at great distances from each other.

Thus, on Leg III, the investigators began to recognize certain formations repeating themselves in different cores and, for simplicity's sake, decided to give them names. In deciding this, they created quite an academic problem for themselves. To the nonspecialist the problem might only seem amusing, but to scientists names are very important business—important for two reasons. To avoid utter confusion, everyone in a field must know exactly what a name means, otherwise communication is impossible. The best example of this is in zoology where each animal has a Latin name that every specialist can recognize. Popularly, the mountain lion, for instance, has half a dozen different names in the United States and Mexico. The other importance of names is a matter of pride. When the *Glomar Challenger* ventured to use the name of the famous original *Challenger,* it was on the dangerous ground of seeming to appropriate other men's achievement. To be allowed to use the name was a great honor.

Naming is so important that in geology, for instance, there is an American Commission on Stratigraphic Nomenclature that must formally approve any new designation. They have a code that suggests, in part, the formal name of a rock-stratigraphic unit, the geographic name "should be the name of a natural or artificial feature at or near which the rock-stratigraphic unit is typically developed. . . . A subsurface may be given a farm name, if its type locality happens to be in some sparsely populated area with few geographic names."

What could be more sparsely populated than the middle of the South Atlantic Ocean? As many scientists thought, it was impossible to follow the Code when designating the Mid-Atlantic Ridge formations; there are few available geographic names there suitable for that purpose.

So they committed what has seemed a heresy to certain other marine geologists. They named their typical formations after famous exploratory ships. There were nine formations needing names and they wanted to arrange them alphabetically, from A on down, with A being the youngest, for ease in remembering. The details were worked out by Dr. James E. Andrews, a geologist from the University of Hawaii who was aboard for the voyage. He named the youngest ooze Albatross, for a Swedish ship that made an epic scientific voyage in the late 1940's; then Blake ooze, after a United States surveying ship that did a great deal of work around the turn of the century; Challenger for the original of that name; Discovery after several British oceanographic ships; Endeavor in honor of Ernest Shackleton's vessel that explored in the Antarctic; Fram for Fridtjof Nansen's ship that survived for almost two years frozen into the Arctic ice; Gazelle and then Grampus for two vessels that not one in a thousand oceanographers ever heard of; and the last was Hirondelle, after the yacht of the Prince of Monaco who was one of the pioneering ocean scientists. Purists are not happy about this breach of the Code, and the problem now is in the hands of the American Commission on Stratigraphic Nomenclature.

On January 24, 1969, the *Glomar Challenger* sailed into the spectacular harbor of Rio de Janeiro after 55 days at sea and after having traveled almost 7,000 miles from Dakar. Thirty-five days had been spent in drilling operations; 17 separate holes were drilled at ten different sites. The drillers had recovered 2,509 feet of core.

The Chief Scientists aboard for Leg III are very certain they proved that the South Atlantic is spreading, and that it has been doing so at a steady rate of a bit less than 1 inch per year for the last 70 million years. They believe that Africa and South America began to separate some time between 130 and 150 million years ago. The continents have not moved apart in a straight east-west line but have rotated a bit. The varying nature of sediments near the ridge crest seems to show that its elevation has varied considerably in the past; often it was much deeper than it is today, which may perhaps be due to a change in the rate of spreading at some very distant time in the past.

Over the years a more detailed history of the South Atlantic will certainly emerge, a more specific date agreed upon for the beginning of the separation, and a better understanding arrived at of the forces that move continents around in this manner.

Yet the great fact arrived at is that continents *are* drifting. If the *Challenger* had never sailed again, after Rio, it would have already earned a great name in scientific history.

10

"During the life of a man, the sea floor moves the length of his body."

—MELVIN N. A. PETERSON,
Chief Scientist of the Deep Sea
Drilling Project

WHEN the *Challenger* left Rio to head north and up into the Caribbean to study its complex set of features, it carried a new scientific staff that was itself complex and colorful.

The University of Miami (one of whose scientific stars, Cesare Emiliani, had done much to get the whole project into motion) was represented by Dr. Richard G. Bader as Chief Scientist, Dr. William W. Hay, Professor of Geology, and Walter B. Charm, a Research Associate. Dr. Bader, an oceanographic geochemist who had been with the University of Hawaii, is Associate Director of the Florida institution, and heavily involved in the administration of the rapidly growing school and research facility a few miles south of downtown Miami.

Robert D. Gerard, Co-Chief Scientist on Leg IV with Dr. Bader, became involved with deep-sea drilling, on a large scale, as senior scientist on the *Caldrill* when it first proved the feasibility of such work to a skeptical United States Congress. From Lamont, Sam Gerard, as he is generally known, was well aware of how important it was for the *Caldrill* project to work out if large-scale drilling was ever to be financed. The ship sailed from Jacksonville loaded with more pipe than it ever carried before. The operating budget was only $250,000 for a thirty-day job. (The *Challenger* spends about one-tenth of that entire amount every day.) The *Caldrill* was only designed to work on the continental shelf, at depths of no more than 600 feet, but Gerard and his group put it to work in 3,000 feet of water. What impressed Washington was that the scientists returned it back to port on time, within the budget, and with cores that at the time were taken from record depths.

Sam Gerard and Dr. Worzel (Leg I) of Lamont were in charge of that school's *Vema* during the long search for the lost submarine *Thresher* down in 8,500 feet of water off Cape Cod. The dramatic story of that sad project does not belong here, but it can be said that it was the *Vema* that found the sunken ship, using equipment that Gerard had improvised and a deep-sea camera developed by Dr. Ewing.

The New York City Police were so impressed with the *Thresher* work that they asked Gerard to bring the *Vema* to the city, about 20 miles down the Hudson from the Lamont dock at Palisades, New York, to help them find the body of the missing Joe Bonanos. This alleged Mafia member had been kidnapped off the New York streets as he was on his way to testify in a crime inquiry and the police thought his body might have been dumped into the East River, possibly in a car or a cement casing that the *Vema's* delicate instruments could detect. This is hardly oceanography, but the powers at Lamont agreed to give it a try. Sam Gerard sailed the *Vema* into the treacherous East River to search for Joe Bonanos. The underwater cameras were useless in the murky water, but mountains of refuse were spotted, including baby carriages and numerous automobiles but not the missing man. He was later found alive and well in Tucson, Arizona.

Sam Gerard made local headlines a few years ago by proposing to dam up Long Island Sound to solve its pollution problems, as well as to produce a number of other benefits, but any joint action in that body of water would need the agreement of so many local jurisdictions that such a proposal, no matter how sensible, could only be a 24-hour sensation. Most recently, Gerard, with Dr. Worzel, has begun a project, now very much in operation, on the Virgin Island of St. Croix that will provide fish farming, refrigeration, air conditioning, electric power, and fresh water all by bringing up nutrient-rich, cold water from a half mile down. Gerard is obviously versatile but he considers himself primarily a physical oceanographer; that is, a specialist in the motions of the sea, particularly ocean currents.

For recreation he likes to sail and maintains a boat on Cape Cod, called the *Fram*, which he manages to use "about one week every other summer." On vacations he likes to rent a sailboat down in the Caribbean islands and sail for a week or so with his whole family. On vacation, while in college, he spent one summer as a roughneck with an oil drilling crew in Wyoming and so he found it easy to talk with the crew on the *Challenger* and they recognized that he knew what he was talking about.

While waiting in Rio to sail, Gerard looked up his old friend, Walter Charm. Charm was a strength enthusiast and introduced Gerard to a Frenchman, a judo expert, who took them to a Japanese restaurant (in Brazil) where they drank saki. Later, Walter Charm wrote, for official publication, *Sea Frontiers*, "My first view of the *Glomar*

Challenger was from the window of a taxi caught up in the unbelievable traffic of Rio de Janeiro on a late Friday afternoon. As we threaded our way through the trucks and buses, the 200 foot drill tower came into view, projecting high above anything in the dockyard. I was expecting a large vessel, but not one of such imposing structure. I thought back to May, 1964, when I boarded a small drilling vessel that was scheduled to drill six holes in the continental shelf off Jacksonville, Florida, one hole to be drilled in the then unheard-of depths of 3,000 feet of water. Because that first venture into deep-water drilling was a success, and recovered enough material to provide a tremendous impetus to the study of the deep basins, I was now about to depart as one of 14 scientists on the fourth Atlantic leg of the Deep Sea Drilling Project."

Also joining the ship in Rio were men from Scripps, Woods Hole, the NSF, an oil company, a geologist from Switzerland, filmmakers Eric Saarinen and Giancarlo Lui, as well as Marcus Aguiar Gorini and Renato O. Kowsmann from Rio. The latter two were marine science students who had visited the *Challenger* in port while an open house was being held. Gerard was so impressed with their understanding of the ship's work and objectives, and their enthusiasm, that he offered them a job aboard as far as San Juan, Puerto Rico. This gesture caused complications with the bureaucracy in Washington and La Jolla. No provision existed to pay them, they did not fit in with the plans, and, international as the objectives were, the youths were not American citizens. Gorini was an assistant professor, but Kowsmann was just an undergraduate. Finally, it was decided that the rules would allow them to work for free and that their plane passage home could be paid for. But there were complications with Brazilian officials as well, and Gerard did not believe they would make it past the red tape. Finally, just five minutes before sailing, the two Brazilians ran down the dock carrying their bags, ready to go to work.

Another passenger, an unofficial one, was not discovered until the second day out. He did not come regularly to meals but stole cheese from the icebox late at night. Finally, on the second night, he gave himself up as a stowaway. The crew had noticed him around but he had such long hair they thought he must be a scientist. Guillermo Batista was a Colombian who had once had working papers as a nonresident in the United States. He had gone home to Colombia and then decided to give Brazil a chance but, seeing a United States ship, thought he would like to see New York again. Batista was also put to work, which he did very willingly. He became so popular with the crew that just before the ship got to San Juan, where he would have to be turned over to the authorities, they took up a collection to pay for his airplane fare home to Colombia.

The first several sites, about 250 miles east of Recife, Brazil, were to be important in that they would sample the basement near the continental margin. The locations resembled the ill-fated Site 8 of Leg II. Then, 12 days out of Rio, the worst accident the *Challenger* has ever had occurred. While the drill pipe was being pulled out of the hole, part of the machinery on the derrick failed and 15,000 feet of pipe tumbled down, lost forever in the deep sea. At a cost of about $10 a foot, this was quite a blow. The official operational summary blamed the accident on a faulty mechanical system, but some people said one of the roughnecks was also at fault; perhaps it was just a moment's inattention. It was decided not to fire him. The man had a good record, he would never let the same thing happen again, and the morale of the drilling team was too important to cause bad feelings over something that was irretrievable. The two drilling crews aboard, each doing a 12-hour watch, were so competitive that haste may have had something to do with the accident. Global Marine Inc. was asked to slow their men down a bit. On the positive side, these first holes did not discover any chert, adding confidence to the statement that the South Atlantic and North Atlantic are distinctly different geologic provinces.

Just about on the Equator a 600 mile long, narrow feature exists known to oceanographers as the Brazilian Ridge. Its origin is unknown at present and drilling was undertaken there, and 180 feet below the ocean floor, a manganese nodule was taken in the core barrel. It had been resting on a thick limestone. The water depth was 6,286 feet. Manganese nodules are the reason for all the stories about great mineral wealth that can be taken from the sea. In addition to manganese, a metal in very short supply in the United States but one vital to hardening steel, the nodules often contain copper, cobalt, nickel, and other valuable metals. Usually they rest on the ocean floor surface, and *Challenger* drilling found no other ones deeply buried. The origin of the nodules is a mystery, but most scientists believe they must be precipitated out of sea water. Often something on the sea floor, such as a shark's tooth, may attract metals in the sea water and, over perhaps millions of years, grow larger and larger. A piece 75 pounds was found in the North Pacific. The nodules are usually found where there is little sediment being deposited at the present time. Many exist on the Blake Plateau, which is usually swept clean by the Gulf Stream.

Mel Peterson has a theory that manganese nodules actually float on the ocean floor. When cut in half, they are often many-layered. They are surprisingly light. Peterson believes that if you sectioned a nodule and then cored the sediments just beneath where it lay, each section down would correspond to a layer on the nodule. To say it another way, the nodule stays on top, acquiring a thin layer of the sediment being deposited, but then, as a new type of sediment is laid down, it floats on top and begins a layer of the next kind of substance. This experiment has not been carried out, but Peterson would like to see it done. It would not need so elaborate a drilling tool as the *Challenger*.

Finding limestone, a shallow-water formation, at a depth of 6,000 feet with a nodule on top suggested to the scientists that the Brazilian Ridge had once been very near sea level but having since subsided to its present depth. They tried to get rocks from beneath the limestone but could get no cores at all. At the point of contact between the nodule and the limestone, they secured a date of about 20 million years ago.

Then, on this leg with its variety of objectives, the *Challenger* moved on to the Vema Fracture Zone, discovered and named by Lamont's hardworking little ship. When Bruce Heezen and Marie Tharp of Lamont published their "physiographic diagrams" of the North and South Atlantic in the beginning of the 1960's, they showed comparatively few fracture zones. When Heezen and Tharp supervised a new diagram of the Atlantic, which the *National Geographic* published in 1968, the back of the ridge was broken by fifty or more fracture zones. Such is the rate of oceanographic discovery in recent years. Many of the fractures look as if blocks of the earth's crust had moved laterally, at right angles to the line of the ridge. Others simply look like ribs attached to a backbone. What causes these fractures to occur is not certain nor is their age clear in relation to the age of the ridge. Some scientists believe that the fractures are extremely old lines of earth motion that have been reactivated as the ridge rises. Quite recent thought is that fractures, often called transform faults, occur because one side of the ridge is sliding against the other. The Vema Fracture is a long narrow trough leading from the ridge westward at about 11° north and the *Challenger* went to sample the sediments in it. They are quite thick and, presumably, of great age. The ridge, of course, is almost bare of sediment and its crest is dated as not much more than 10 million years old.

Coring was done down to 1,586 feet and nothing older than the Pleistocene (Ice Age), about 500,000 years, could be found. It seemed that the sediments, from their contents, came from the Amazon River. They contained gem minerals native to Brazil, such as aquamarine, topaz, and tourmaline, and layers of plant material up to one inch thick, which must almost surely have come from large Amazon floods. This material had been transported halfway across the Atlantic.

The paleontology of the cores suggested that sediments were forming at the rate of one foot every 250 years, a very rapid rate. From the evidence of the cores, the scientists suggested that the Vema Fracture Zone could be quite old, Tertiary, older than the ridge crest, but that it did not begin to fill with sediment until the broad Demerara Abyssal Plain, between the fracture and the continent, had been filled. The other explanation, which they liked better, is that the fracture is Pleistocene, no older than one or two million years, and that it is being rapidly filled with sediment as it forms. (The dating at this site was done on the basis of estimates worked out by Cesare Emiliani of Miami for other oceanic cores.)

Although the *Challenger*'s drilling on Leg IV did not solve all the problems in world geophysics, its course, carefully planned, naturally, took it near some of the biggest questions. Three of the questions are: Where has the Caribbean been in ages past? Where was it before the Atlantic opened up? Why do island chains form in arcs?

A glance at a map of the Caribbean shows the islands to the east of Puerto Rico forming a very definite curve as they progress step by step south to Trinidad. The Pacific has many such curving chains of islands. Many of these island arcs have deep trenches that lie on the outside of the curve. In the West Indies, the trench is north of Puerto Rico. At 30,180 feet it is the deepest spot so far discovered in the Atlantic. Many island arcs have volcanoes on the inside of the curve, and the West Indies arc has a number of volcanoes, some fairly active and others extinct. The similarity of these island chains certainly has some significance. As the *Challenger* approached this West Indies arc, it first stopped to sample the ocean floor 200 miles north of Barbados. Barbados itself is definitely not part of the West Indian chain. It is nearly two hundred miles east of the arc and is not volcanic as the other islands are. In addition to whatever light drilling here might shed on the causes of island arcs and, as always, on sea-floor spreading, it was hoped that the sediments might show patterns of fossil animals so that comparisons might be made between the Caribbean and Mediterranean.

What the drilling produced was a hard clay that was exactly like a formation found on Barbados itself. The core was about 1,500 feet below the ocean floor. On Barbados, the same strata are known as the Oceanic Formation. It has been famous for its well-preserved and varied open ocean microfossils. The formation is exposed on the east slopes of Mount Hillaby. The age is Eocene, about 45 million years ago. The demonstration is that Barbados is a block of the deep ocean floor that was raised in time from a depth of three miles. That this is not obvious to every visitor to Barbados is due to the fact that much of it became covered with coral as it rose toward sea level. What might raise the ocean floor three miles, in this region, is not yet perfectly clear.

After a 36-hour stop in San Juan to offload the super-cargoes from South America and pick up Archie Mc-Lerran, the drilling expert, the ship went to a position north of the Puerto Rico Trench. The site was on the outer ridge of the wall. Surveys had shown two reflectors and the desire was to find out if they were part of Horizons A and B that cover so much of the western Atlantic. Another part of the Puerto Rican Trench problem is that the tiny fossils found in the sediments are shallow water types: Why are they found at such depths? Drilling was very unsatisfactory here, the bit destroying itself on the chert. Then three sites within the Caribbean were covered to get the oldest sediments possible within an island arc, to get a complete stratigraphic record of the sediments, which are thick here, and to discover the age of the seismic reflectors, which resemble Horizons A and B out in the Atlantic. The cores on the first site, surprisingly, showed an absence of sediment deposition for 20 million years, but a very long section of pure radiolarian ooze was obtained, covering a period of 5 to 10 million years. During this time the microscopic animals underwent profound evolutionary changes, and these can be correlated from deposits as far away as Israel and the Central Pacific. The horizons had the same age as those in the Atlantic.

One of the sites was near the Aves Ridge, a shallow feature west of the little Grenadine Islands that Bruce Heezen believes to be continental granite. (A French fleet on its way to conquer Jamaica piled up on the Aves Ridge in the seventeenth century.) The cores here showed a great deal of volcanic material from the Pleistocene, indicating a good amount of volcanic activity, presumably from the islands to the east, about a million years ago.

The drilling in and around the Caribbean leaves that ocean, or sea, very much of an enigma; still it seems likely its history is much more complicated than that of the Atlantic.

The *Challenger* reached Cristobal in the Panama Canal Zone on March 4, and the scientists left the ship. Archie McLerran boarded a plane for Dallas where he planned to change and fly to his home in San Diego. As a footnote to colorful Leg IV, his aircraft was hijacked and he found himself in Havana. The passengers were treated well there, however, and in time all were taken safely to Miami. In Miami, though, the customs officials harried the poor innocent people trying to find out if they had picked up any Cuban cigars or rum and were smuggling them in. The FBI intervened to stop this harassment, and McLerran finally reached San Diego and his worried wife. The *Challenger* then passed through the Canal and into the Pacific for the first time. Then it sailed to La Jolla with its Scripps Institution and was given a joyous welcome.

For visitors to the ship during its first trip to California the officials wrote a pamphlet explaining some of the highlights of the great venture. It read, in part, "With the completion of drilling and coring operations in the Atlantic Ocean, the DSDP has reached the halfway mark with eloquent results. When the drilling ship *Glomar*

Challenger steams out of San Diego, five 55-day legs remain in the Pacific Ocean.

"Deep-sea drilling is now a very successful reality. Many miles of actual penetration, about 4,000 feet of core recovered so far, maximum penetration of 2,738 feet and maximum drill string length of 19,075 feet, so far, show that drilling in the deep oceans is a remarkably potent tool for the study of marine geology.

"Scientific and economic results thus far completely bear out our great expectations. Theories of sea-floor spreading away from mid-ocean ridges have been very strongly substantiated, and rates can now be measured from actual samples and compared with evidence from pole-magnetism. The igneous rock underlying the sediments, i.e., the initial floor of the ocean at any one place on which sedimentation began, has been sampled at many places in both the North and South Atlantic and its immediate origin elucidated. Evidence for major, oceanwide changes in sedimentation, surely related to major changes in oceanic circulation and chemistry, has been found. Reflecting horizons in the sediments have been sampled and identified. Deposits by turbid flows, reaching far out to sea, have been shown to be of much greater importance than heretofore suspected. Eocene chert is exceptionally widespread in the Eastern and Western Atlantic.

"Salt domes, similar in almost all regards except depth of water to notable oil-producing structures of the Gulf coast of the United States, have been proven to exist in the Gulf of Mexico. These are the Sigsbee Knolls, and the discovery of oil in the cap rock of the knoll that was drilled is a distinct milestone in the understanding of formation and accumulation of petroleum in deep oceanic areas. It is far more than just the discovery of a potential new oil province; it is a new direction of thought."

11

"Mars never got big enough to have much geology."
—Sir Edward Bullard,
Eminent British geophysicist

IN 1969, when the *Glomar Challenger* first entered the Pacific, many scientists trying to understand the nature of oceans and continents were being strongly influenced in their thinking by a theory first appearing in print the year before. Under the name of "plate tectonics," this theory is closely associated with the names of W. Jason Morgan and Xavier Le Pichon. Their concept of the world is that large areas of it are divided into great plates, perhaps ten of them, which move in separate, rigid crustal blocks. The major reshaping of the earth's crust, the tectonic activity, is all occurring where these plates collide.

(A cautionary note might be made here saying that the plate theory is so new that opinions concerning it, the number of plates and their various motions, are still in a state of flux and often change rapidly.)

The depth of the plates is unknown but the present best estimate is that they are somewhere around 75 miles thick. Thus, they reach into the earth's upper mantle and a continent and a vast amount of ocean floor may both rest on the same plate. As presently conceived, the major plates are moving in directions that of necessity cause collisions. One plate consists of all North and South America and the Atlantic up to the Mid-Ocean Ridge. Its western boundaries are the Pacific coasts of the two continents. The motion of this American plate is westward. The Caribbean is a separate plate that has been bypassed. Although North and South America are joined now they may have been distinct and moving free of each other 100 million, or so, years ago.

The African plate begins just east of the Mid-Atlantic Ridge, extends into the Indian Ocean, and stops at the Carlsberg Ridge. The southern boundary of the African plate is unclear, but it extends far south of the continent. To the north, it stops at the Mediterranean shore on the west, but to the east it crosses the sea and includes parts of Greece and Yugoslavia. Farther east it is bounded on the Red Sea. Across this narrow water, which is certainly a part of the rift, there is a small plate that is Arabia.

The Eurasian plate begins at the Mid-Atlantic Ridge and includes all of Europe and Asia, with the exception of India. It extends out into the Pacific, probably up to the line of deep-sea trenches on the western edge, but does not include the sea in the Pacific east of the Philippines. The Philippine Sea is another small plate by itself. The Eurasian plate is moving eastward and slightly south.

The Indian plate includes a considerable amount of ocean floor to the south and this large segment of the earth is moving north, pushing in on the Eurasian plate. Australia is another plate, also moving north. The Antarctic is still another individual plate.

The enormous Pacific plate is bounded by the great system of deep trenches that ring the ocean. Its southern boundary is marked by the East Pacific Rise, the Pacific counterpart of the Mid-Atlantic Ridge. Where the Pacific plate touches North America its boundary is the San Andreas Fault, an easily understood example of tectonic activity. Off British Columbia and Southern Alaska there is a small, remnant plate, which is yet unnamed and whose motion is unknown. It is not certain where the Pacific, the American, and the Eurasian plates make a triangular junction around Alaska or Siberia, but the understanding is that they do not meet in the quiet, shallow Bering Sea. The Pacific plate's motion is to the northwest.

As the East Pacific Rise, the eastern boundary of the Pacific plate, moves north it passes close to Easter Island. The sea floor between this rise and South America is another plate and it is moving to the east, into the South American continent. The northern part of this unnamed plate is moving in a somewhat different direction, to the northwest, and may be a distinct plate by itself as some recent research seems to indicate. This comparatively small plate may be the same as that of the Caribbean, even though Central America rises between them. As may be noted, the Caribbean produces not only revolutions and hurricanes but also a great deal of trouble for geophysicists.

This marvellous new idea of plates fits in very well with the concepts of sea-floor spreading and continental drift. It can also explain the great mountain ranges of the world, deep-sea trenches, vulcanism, and earthquakes. Seldom can a mint-fresh idea have been accepted so rapidly, even as a working hypothesis, by members of any scientific discipline.

Practically, why does the theory of plate tectonics help understanding so much? What happens when an ocean meets a continent? In the South Atlantic, for one case, nothing happens at all. No earthquakes or any other great activity occurs because the eastern edge of South America and the western edge of Africa are each part of the same plate as their adjoining ocean floors and the plates move as units. On the western side of South America, the meeting of an ocean and a continent produce tremendous results because here two plates are moving in opposite directions and something must give. What happens is that the heavier material of the ocean plate is forced under the lighter continental material, while some of the light sediments from the ocean floor pile up on the landward side of the trench to build the mountains. The point of collision is a deep trench, almost at the shore's edge, which is the reason western South America has essentially no continental shelf. As the heavier oceanic material descends, it causes the notorious South American earthquakes. At a certain depth this material is presumed to melt under the heat of the earth and some of this material, reworked with rocks already existing there, is brought to the surface again as granite in the mountains and volcanoes of the Andes.

Sometimes a plate may have a chain of islands at its border and, in this case, the colliding, sinking plate is also forced down into the earth. This occurs in the Aleutians, south of Indonesia, off the Tonga Islands, in the Caribbean, and other places. Here a trench is formed to seaward, where the plate is descending and volcanoes rise on the islands themselves. Deep earthquakes occur in enclosed seas behind the island arcs. The Sea of Japan is a good example of this.

Plate tectonics offers an explanation of why the oceans have no sediments older than 200 million years. The oceans were certainly there, but gradually they have descended into the earth at the trenches or, in some other cases, piled up into mountain ranges at the point of collision. Thus, when two continents collided, as when India struck Asia, both plates were made of light rocks so that they jumbled up together and produced the Himalayas, where deep-sea fossils have been found in peaks higher than 20,000 feet.

This theory of plate motion also offers an explanation for the old Appalachian mountains of eastern United States and their counterpart mountains in northwestern Europe. Presumably, if the great plates are in motion they did not suddenly begin with the splitting of the Atlantic 200 million years ago but have always been moving about. Perhaps, in some previous collision of an older Atlantic, the Appalachian-northwest Europe range was formed (much higher than it is today) and then was separated when the Atlantic split apart again. Similarly, the Ural Mountains between Russia and Siberia, unusual because they are in the middle of such a vast continent, may be understood with the idea that what is now Eurasia was once two plates that collided at some time in the distant past. The Rocky Mountains, according to the theory, arose when the American plate collided with the Pacific plate and rode over it.

For the Pacific, evidence of a plate can be found, among many other areas, in the Gulf of California. Various kinds of evidence show that Baja or Lower California has been moving away from Mexico proper for 4 million years. The rift that is a continuation of the East Pacific Rise runs between them. Baja California is on the Pacific plate, while Mexico is on the American plate. California, west of the San Andreas Fault, is quite demonstrably moving north and west, away from the rest of the region. Thus California is right on the division between two plates. The westward moving American plate has overridden the East Pacific Rise here, forming the mountain ranges of the state. An interesting corollary of this is the recent suggestion that sea-floor spreading south of the Mendocino Fracture Zone stopped after the East Pacific rise was destroyed in this region by a collision with the American plate, but it has continued farther north. It was as the rise disappeared to the south that the San Andreas system became fully developed.

The same kind of studies that showed the existence of the Mid-Atlantic Ridge show that the break in the Pacific comes in this East Pacific Rise, which disappears into the United States through the Gulf of California and reappears off the coast of Oregon. The same kind of studies show that this is the point of deep-sea spreading. Where, then, should the oldest parts of the sea floor be? Near the continents, as in the Atlantic. The Pacific rise, however, is in the eastern edge of the ocean and the oldest parts of the sea floor there have been consumed under the advancing American plate. The oldest remaining Pacific floor now existing must therefore be at the opposite side. Since the Pacific plate is moving in a northwest direction (as is its eastern edge, the San Andreas Fault), the opposite side must be in the northwest Pacific, somewhere to seaward of Japan, Guam, or the Mariana Islands. As the *Challenger* began to work in the Pacific, it had a number of objectives but the most important were to test the spreading idea and find the oldest possible sediments. The concept of the world divided into vast plates makes the choice of drilling sites a great deal clearer.

12

DEEP DRILLING PSALM

Mel Peterson is our leader; we shall not want.

He does not maketh our bed, he leadeth us to the deep water.

He restoreth our souls; he leadeth us in the track of *Argo*
    for DSDP's sake.

Yea, though we shall drill through the valley of opaque sediments,
we will fear no twist-offs; for our chief scientists are with us,
    their plans and profiles they comfort us.

They preparest site surveys for us in the presence of our enemies—
    the cherts; they fill our vans with cores, our notebooks run-
    neth over.

Surely, press conferences and grants shall follow us all the days
    of the project; and we will report on the results forever.

—Attributed to DAVID BUKRY, Leg VI
(Posted in Science Lab, *Glomar Challenger*)

THE FIRST cruise out in the Pacific, numbered Leg V, had a very worthwhile plan, with two major objectives. The ship would sail north from San Diego along the California coast until it reached the Cape Mendocino area where the San Andreas Fault enters the sea and where the Mendocino Fracture runs west along the ocean floor for thousands of miles. The ocean floor off California has a number of great fractures, all spaced rather equally distant from each other, and all at right angles to the East Pacific Rise, which is thought to lie here under the continent (as the San Andreas Fault). These great fractures were discovered by their magnetic anomalies, anomalies whose patterns are similar to those mapped in the Atlantic. The hope was that the Pacific-California anomalies could be dated by their sediments, as they were in the South Atlantic and that some orderly, worldwide scheme would emerge.

As the second objective, the *Challenger* would then head out to sea and the longitude of 140 degrees west. Drilling then would be carried on due south, along 140 degrees west, until the ship was only 15 degrees above the Equator. The line of longitude here is thought to be the axis of rotation of the Pacific plate. Two ocean currents were involved in the calculations for this course. One, the North Pacific Gyral, closely resembled the currents in the North Atlantic where the Gulf Stream is the best-known feature of the circular pattern of motion. The other current is the Equatorial that runs westward across the Pacific, on the surface, under the influence of the prevailing winds. At present, animal and plant production at the Pacific Equator is much greater than it is farther north. Thus, if, as supposed, the Pacific plate is moving to the north, the old Equator floor is also farther to the north and sediments here should show much greater productivity at some time in the past. Sediments at the present Equator should show that the Pacific was much less productive there when the sea floor was far to the south. This part of the cruise would also provide reference sections of microscopic fossils from depths that the paleontologists had not sampled in this part of the ocean. After about 55 days at sea, the cruise would terminate in Honolulu.

Under the scientific direction of Dr. Dean A. Mc-Manus, of the University of Washington at Seattle, and

Dr. Robert E. Burns, oceanographer in charge of a Federal research project at the same school, the *Challenger* left San Diego on April 12. Some, but not all, of the sites had been studied in advance by the Scripps research ship, *Argo*.

No one can be faulted that Leg V did not do much to advance geophysical theory. The broad stretch of the Pacific covered was not one that, with the tools then available, would yield much to drilling. The basalt usually came far too close to the surface; sediments were seldom very thick and often there were few informative fossils, too often just clays or mud. Chert stopped the drilling in a number of cases. When sediments could be sampled and dated where they met the basement, the age agreed with that predicted by the magnetic anomalies, but at one site there was a discrepancy of 15 million years. The scientists decided that either the oldest sediments had not been preserved or a flow of basalt had buried them. In their report, Drs. McManus and Burns strongly urged that a reentry system be worked out to go back into a partly drilled hole so that the bit could be changed in "impenetrable" chert.

A typhoon that came up on the next cruise of the *Challenger* might be expected to have had a dampening effect. The ship's weather satellite receiver was supposed to warn of such things but this storm developed right overhead while the vessel was between the islands of Yap and Palau in the far western sea, a birthplace of these great Pacific storms. Very soon winds of 80 miles an hour and 65-foot seas developed, but position was held with the bow thrusters and the main propellers. A bit of weather did not deter Leg VI.

During this drilling, which began in Hawaii and ended

up in Guam, the *Challenger* had two eminent Russian scientists aboard, A. P. Lisitzin and V. Krasheninnkov of the Academy of Sciences in Moscow. The latter could speak only a very formal English but Lisitzin could manage a quite colloquial brand of the language. As told by Dr. Tony Pimm, a young scientist from Scripps, some of the more playful scientists and technicians aboard began to use the ship's public address system in the style made famous as the U. S. Air Force flyer's pattern for voice communication. Thus, to end a conversation you would say "Roger and Out." A call might come over the PA system in this way. "New core coming aboard. Looks like Miocene. Is there a paleontologist available?" Someone would answer. "Paleontologist on his way. Over, Roger and Out." Dr. Lisitzin listened to this kind of talk for a few days and then asked Tony very confidentially, "Tell me. Just who is this Roger?"

Relations were excellent between the nationalities and, like everyone else, the Russians were properly amazed by the quantity and quality of the food. On the day the first astronauts landed on the moon, the *Challenger* people followed events over their radio and when they heard the words "The Eagle has landed," the Russians cheered as much as anyone else. Then Lisitzin went to his cabin and returned with two bottles of Russian vodka and all hands toasted the event. On the Fourth of July, the cook baked a birthday cake for the United States and the Russian scientists applauded with all the rest.

The Chief Scientists aboard were Dr. Alfred Fischer of Princeton who had sailed on Leg I and Dr. Bruce Heezen of Lamont-Doherty. Fischer's degree was in paleontology, but only the specialists can tell where this science becomes geology. Geologists sometimes innocently think in terms of geophysics and Fischer worked for a number of years as an oil geologist. Then he went to Princeton as Professor of Geology.

Heezen has been mentioned for his work on the Mid-Atlantic Ridge and his physiographic diagrams of the ocean floors and has been called "a major painter of undersea geography." He has said "I don't fiddle around in shallow water" and he once spoke of the drilling project as "a bold sort of adventure like this." He prefers to think in global terms.

The biggest story to come out of Leg VI must be the overall view it gave of the North Pacific Ocean's history. The drilling between San Diego and Hawaii had shown, in a general way, that sediments become older as you move westward in the ocean. Continuing on toward Asia, *Challenger* cores showed continuously increasing antiquity until in the northwestern Pacific, east of the Mariana-Bonin Islands and north of the Caroline Islands, an area south and east of Japan, a remnant of Early Mesozoic ocean floor was discovered. It had an age in excess of 140 million years and is thus the oldest sediment found so far in the Pacific. Heezen, Fischer, and their associates were able to draw a map, based on their sediments, showing a curving pattern of successive ocean floor ages all the way across the sea from the East Pacific Rise. Just west of the Rise, the sediments were merely Pliocene-Pleistocene, 1 to 13 million years old; then an Oligocene region, 25 to 36 million years old; and Eocene-Paleocene 36 to 63 million years old; and so forth, until at the apex of a triangle formed by these increasingly smaller circles the predicted Mesozoic (Jurassic) sediments were found in the proper place.

The obvious question is, if you can find this nice symmetry on the western side of the rise, can you find it on

the east? It is possible, but only until you get to an age
of, at the most, 36 million years and then you reach the
coast of Central America. As Heezen comments, "A basic
problem remains about the Pacific. There is no symmetry.
Either the ocean is asymmetrical, which no one likes, or
somehow you must account for the destruction of 6,000
miles of ocean floor."

A projection, based on the assumption that the advancing American continents destroyed this ocean floor, would
place the counterpart of the oldest part of the Pacific
somewhere far out in the Atlantic. But this has not yet
been discovered. According to Heezen, "An overall picture of the past is not yet possible."

An assumption easy to make when hearing about this
big picture is that North America is getting closer to
Asia. Heezen thinks not, but that new ocean floor is
being created west of the East Pacific Rise and this is
going down into the deep trenches on the west. But
then, to confound the layman, he asks, "Is the floor
really being consumed in the trenches? How can we
measure such a thing?"

Heezen is not convinced that the oldest part of
the Pacific has really yet been found. He thinks something a bit older may be discovered and he will be on
the cruise of the *Challenger* when it returns to this area
to drill again somewhat nearer to Japan. He believes
that the Pacific may possibly be a little bit older than
the Atlantic—but not to any great extent.

There is a considerable amount of Pacific Ocean between these oldest sediments found so far and the Asian
continent itself. It would be reasonable to suppose that,
if one continued west in this direction, more aged seas
could be found. This is not the case, however. The Mariana Trench, the deepest in the world known at present,

lies not too distant to the west of this historic drilling, and beyond it, in the Philippine Sea, the deepest sediments are suddenly a great deal younger.

As Tony Pimm, a scientist on Leg VI, wrote concerning the Philippine and similar seas, "How can such large areas exist on the landward side of these trenches?" [These trenches are assumed to destroy the spreading ocean floor.] "The rough topography and thin sediment cover revealed by various survey ships indicated that these seas are comparatively young. Leg VI of the Project drilled two holes in the Philippine Sea and discovered that the oldest sediments here are of uppermost Oligocene age (30 million years). This basin has, therefore, had a much shorter history than the neighboring ocean."

A possible explanation of this is that west of the Mariana Trench, in the Philippine Sea, a sort of mini-Mid-Ocean Ridge has begun and is operating independently, but in the same way, as the larger worldwide system. The Mariana Trench does have eruptions along its axis, shallow earthquakes and high heat flow, just like the Mid-Atlantic Ridge.

Another possibility to explain the young floor of the Philippine Sea is that the East Pacific Rise once circled the entire Pacific, including the western and northern sides and that somehow it disappeared. This was rather difficult for the scientists on Leg VI to accept. In an official report they wrote, "The data presently available to us do not permit a resolution of this problem. In any case, the association of young crust with an island arc and trench system that appears to have been in existence since Eocene times (perhaps 50 million years) or longer suggests that current concepts of the mechanics of island arcs are too simple." It has been said that, in science, the moment you find out something new you also create a new question.

On a slightly less cosmic scale, drilling in the area around the Caroline Islands showed some large areas of submarine lava flows that did not produce volcanic peaks but a very flat, smooth ocean floor. Such flows resemble the lava plains of Oregon, Idaho, and Washington and conceivably have covered up older layers of sediments. Drilling south of the Hawaiian Islands and on a large guyot (a flat-topped undersea volcanic peak) named the "Horizon" was interrupted—in both cases by chert. They speculated, as other drillers did, on its age and mode of origin and considered the question filled with "much uncertainty." It did seem to them that silica-bearing skeletons of diatoms, radiolaria, and sponges turned into opaline and remained in this condition for as much as 25 million years. Opaline skeletons and chert do not occur at the same levels. At the beginning of the Eocene epoch, massive chert beds begin to occur and opaline matter disappears. Most of the chert in the North Atlantic and Caribbean was also dated as Eocene.

Although the drilling did not show much about seamounts, and guyots, Bruce Heezen remarks that they are all at least 100 million years old and that paleomagnetic dating shows that they were formed in latitudes near the Equator and gradually moved north, perhaps with a northward motion of the Pacific crust. They may also have a westward motion, which would agree with the plate tectonics theory, because they become deeper as they progress to the west. Heezen also points out that there is a progressive subsiding toward the west in the Hawaiian Islands. The highest and largest is Mauna Kea, the easternmost of the Hawaiian range, while the smallest and lowest is Midway, the island farthest to the west. Unfortunately, Leg VI drilling did not do much to add to these certainly plausible ideas.

Two and a half weeks from the end of the trip, the *Challenger* ran out of the beacons necessary for dynamic positioning. At considerable expense a ship was dispatched to bring out new ones. This was not due to lack of foresight in carrying enough beacons but to the fact that a number of beacons failed and the fact that an unusually large number of holes were drilled. How is one to balance the costs of such emergencies? The official decision seems to have been that once such an expensive ship is at its desired location, you find, if you must, the necessary money to keep it there.

When the *Challenger* put into the harbor of Agana on Guam, the scientists all were flown to San Diego. At a press conference there, Dr. Lisitzin spoke with reporters. "If I were a capitalist, I would not hesitate to put my last shirt on the ocean." Because the moon samples brought back by the United States astronauts showed no signs of life, Dr. Lisitzin said, "This event [the drilling] must be of more importance to man. There must be something here very useful to humanity. The commercial ocean can offer to humanity things we can not even imagine. The simple idea of the ocean as a supply of fish is at an end, and a new vista is being opened that can offer more in more varied ways. The ocean is like a hidden treasure."

The average scientist on a *Challenger* cruise may take a few samples of sediment off the ship for immediate work back in his own laboratories. The Russians took about 500 samples. They filled four large boxes. Out of courtesy, no one said anything to the Russian guests about this, but no one publicized the matter either, out

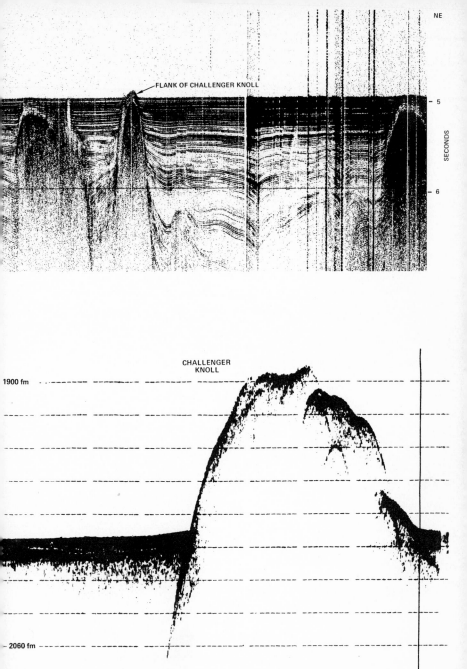

NE

FLANK OF CHALLENGER KNOLL

— 5

SECONDS

— 6

CHALLENGER
KNOLL

1900 fm

2060 fm

examples of undersea exploration by sonar techniques: Top: A profiler traverse of the Challenger Knoll area in the Gulf of Mexico, off the Yucatan Peninsula. *National Science Foundation*. Bottom: An echo sounder traverse of the Challenger Knoll. *National Science Foundation*

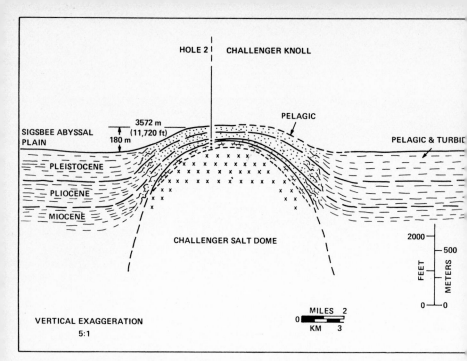

Above: A drawing of the Challenger Knoll, using the information gathered by the *Glomar Challenger*. *National Science Foundation*

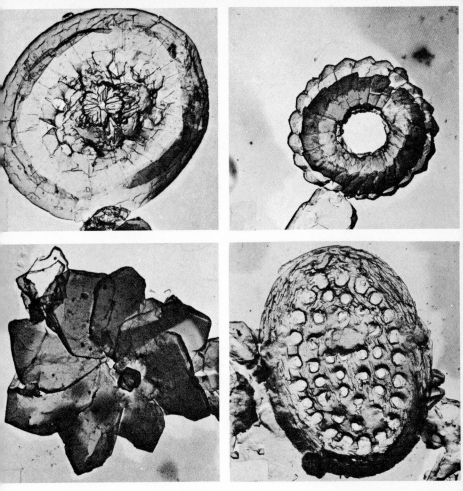

Cross sections of various cores, magnified 10,000 times, showing the microscopic fossils that are so important in revealing the ages of the oceans and the drifting of the continents. *National Science Foundation*

Above: This 170-million-year-old li
stone core, the oldest ever taken f
any ocean, was recovered at a dril
site near the Bahamas. Holding the
toric core on the drill floor of the *Glo
Challenger* are Dr. Charles D. Hollis
John I. Ewing, and James A. Dawson
Gulf Oil Co. *Scripps Institute of Oce
ography*

Left: Dr. Maria Cita, a paleontologist f
the University of Milan, examines s
ments from a core just brought up to
drilling floor of the *Glomar Challen*
Dr. Cita was one of several women sc
tists who joined the ship on its vari
trips. *Scripps Institute of Oceanogra*

wo prominent Russian scientists, Dr. Valeriy A. Krasheninnikov, second from left, nd Dr. Alexander P. Lisitzin, second from right, were members of the scientific team n the *Glomar Challenger*'s Leg VI. They are seen here with Dr. Bruce C. Heezen, left, nd Dr. Alfred G. Fischer. *Scripps Institute of Oceanography*

Above: The reentry cone b
lowered over the port side o
Glomar Challenger. It is keel-h
to the opening (the "moon p
in the middle of the *Glomar*
lenger, directly beneath the dr
derrick, and then lowered wit
drill string. The three sonar
cons are visible at the top. *So*
Institute of Oceanography.
The drilling crew rigs equipme
begin drilling and coring o
tions. *Global Marine, Inc.*

ove: Crewmen bring a ninety-
ot section from the automatic
ck to be connected to the drill
e. *Scripps Institute of Oceanog-
hy*. Right: A night shot of the
omar Challenger's drilling rig
or. *Scripps Institute of Ocean-
raphy*

GEOLOGICAL ERAS
(From Present Back to Formation of Earth)*

Years Ago (in millions)	Period of Time	
Present — 2	Recent & Pleistocene	
2 — 7	Pliocene	} Neogene
7 — 26	Miocene	
26 — 38	Oligocene	
38 — 54	Eocene	} Paleogene
54 — 65	Paleocene	
65 — 100	Upper (Late) Cretaceous	
100 — 136	Lower (Early) Cretaceous	
136 — 193	Jurassic	} Mesozoic
193 — 225	Triassic	
225 — 280	Permian	
280 — 325	Upper (Late) Carboniferous	
325 — 345	Lower (Early) Carboniferous	
345 — 395	Devonian	} Paleozoic
395 — 435	Silurian	
435 — 500	Ordovician	
500 — 570	Cambrian	
570 —	Precambrian	

*After R. M. C. Eagar, *The Geological Column*, Manchester Museum, 1968.

of fear of criticism from professional anti-Communists. The Russians in their laboratories in Moscow, however, ran about 100 tests on each sample—a far more complete analysis than Americans would have done—and they submitted hundreds of manuscript pages with their data for inclusion in the official volume on Leg VI. The DSDP people consider it permissible to mention the matter now and they have requested Russian scientists for future cruises.

Mel Peterson believes that the Russians would welcome the *Challenger* to do deep drilling in the Black Sea. That water is so full of hydrogen sulfide, however (it is anaerobic, without oxygen, and fish cannot live in it very far below the surface), that no one knows if the fine steel in the drill pipe could survive the contact. Tests are now being made to discover what exactly hydrogen sulfide would do.

Continuing the exploration of the western Pacific on a track farther south than that of Leg VI, the *Challenger* set out from Guam, destination Honolulu, with a brand-new scientific staff. The lead men, Dr. Edward L. Winterer and William R. Riedel, were both from Scripps.

Dr. Betty Gealy from Scripps was also aboard. On the DSDP her title was Executive Staff Geologist. A Phi Beta Kappa who received her doctorate from Radcliffe, she is the mother of four. Her philosophy concerning the drilling project was very clear and concise. "If we were certain of what we were going to find, we would never need to drill. It is the prospect of discovering something really startling that will radically change our previous ideas concerning the origins and history of the ocean basins that spurs us into the unknown.

"In order to ask questions of the unknown as wisely as possible, however, we formulate working hypotheses, ready to alter these in the face of new information acquired by our search. The drilling program has been designed to test several of these."

One objective of Leg VII was to test the hypothesis that the oldest part of the Pacific might be in the Solomon Rise, a plateau across a deep trench from the Bismarck Archipelago. Another hypothesis to be tested was that there had once been a great ocean rise in the west Pacific comparable to the East Pacific Rise. Named the Darwin Rise by its sponsors, the theory was that it had become extinct and foundered, leaving behind it the chains of sunken atolls and guyots, the flat-topped volcanoes that are so frequent in this region. Last on the agenda was a site on the Hawaiian Ridge, north of the islands, which had been selected as the ideal first Mohole a few years ago.

One more general aim of the trip was simply to examine the age and history of development of the island arc system, the ridges and small basins of the western Pacific, and the age of the central part of the Pacific Ocean basin. Dating the sediment levels was another general goal.

The drilling south of Guam toward the Bismarck Archipelago and New Guinea did not show this region to be the oldest part of the Pacific but comparatively quite young, only about 25 to 30 million years old. This seemed to be caused by extensive volcanic activity that had covered up older sediments, and the investigators recommended more research to understand the cause of all this activity.

Where the Darwin Rise was supposed to be, the sediments showed that the region had been deeper than the depth of carbonate compensation (more than 12,000 feet deep in this region) for at least the last 80 million years.

This results "strains" the concept of a rise. On Leg VI, drilling farther north in the same hypothetical formation, the reflecting layers were found not to be the turbidities that had supposedly tumbled, like landslides, from the sides of the rise. They were typical deep-sea sediments. The Leg VI scientists wrote, "Thus, the Darwin Rise faces abandonment as a useful working hypothesis." This is the polite way in which scientists verbally demolish each other's ideas.

The Darwin Rise idea was based on the fact that in this area many atolls and guyots had sunk far beneath the sea surface. The thought was that the entire sea floor, with atolls and guyots, had sunk more than a mile as a single unit. Leg VII drilling "suggests that the subsidence of guyots and atolls may not be matched by similar amounts of subsidence in the adjacent areas." In another report, "Perhaps some segments of the sea floor move up and down more or less independently, and are not susceptible to a simple, large-scale explanation. The history of changes of depth of the sea floor deserves closer study."

While drilling south of Guam, the researchers found that the warm equatorial current that was usual for the region had been diverted for a time about 5 million years ago. During this diversion, abnormally cool waters flowed across the area and brought in tiny plants and animals characteristic of such temperatures. A new question thus arises: Was this change in temperature due to a change in the arrangement of island barriers, a change in climate or some other condition?

During this cruise, nearly continuous core samples were obtained—continuous cores are extremely rare—and sediment layers rich in fossils that lived over the past 20 million years are now available. Their evolutionary changes will make it possible to produce much more precise paleontological time tables than were ever available before.

The original Mohole site had been surveyed in advance by the *Argo* but, "The location is not well suited for drilling and the Pacific Advisory Panel subsequently approved shifting the site less than 100 miles to the northwest [of Oahu]." The bit was stopped by chert after about 200 feet of drilling beneath the sea floor, about a third of the way to the basement, and the sediments were dated as early Eocene. By extrapolation, this ought to make the regional basement about the right age for the region, as predicted by the overall Pacific theory. Summing up their work in a preliminary war, the Leg VII scientists felt they had "paved the way for future detailed geological exploration of the western Tropical Pacific. In addition, the geologists aboard the drilling vessel *Glomar Challenger* have made several discoveries that do not fit into previous conceptions of the history of that area." At a minimum, these scientists from Scripps had demolished the Darwin Rise, an idea which had originally been developed at Scripps.

13

"... in times to come, these twenty years when men gained a new level of understanding of their planetary home will be thought of as one of the great ages of exploration. ... The simplicity and grandeur of this hypothesis of sea-floor spreading has caught the imagination of scientists throughout the world."

—DR. ROGER REVELLE,
Prominent American oceanographer

AS THE *Challenger* sought the ocean's secrets east and south of the Hawaiian Islands on Leg VIII, it began to near the end of its allotted 18 months and the $12.6 million provided by the United States government. Only a few months remained in which to probe all the rest of the vast Pacific. Of course, the ship's science staff was not interested in mile by square mile mapping, finding new seamounts and guyots, more fractures, and more soundings. These matters would continue to be taken care of by conventional vessels. The *Challenger* objective was to understand the greater picture, the changes over incredible amounts of time, the forces behind such changes. The few holes drilled at great expense might seem a bit silly in relation to the size of the ocean; but these holes were chosen on the basis of considerable past experience and they could, and did, reveal the history of regions the size of entire nations.

The scientific prospectus for Leg VIII began, "The

vessel will head southeast from Hawaii to a site on the 140° meridian, turn due south to 30° latitude and then head for Tahiti." The ship would drill six holes on . a straight line, from north to south, 1,300 miles long. The object would be to find out how Pacific Ocean current systems have varied in the past. It was known that today more sediments are being deposited along the highly productive track of the Equatorial Current than in the less productive waters to the north and south. If the sea floor has remained stationary during the past ages, then the sediments should be comparatively thin at the first hole, become thicker as the Equator is approached, then thin out again as the *Challenger* moved south from the Equator. If, however, the sea floor has moved because of sea-floor spreading then the sediments would be more complicated, but interpretation ought to show such a motion. A complication of the matter is that perhaps the ocean floor has moved laterally, either east or west, due to fractures on the north-south line. This seems a remarkably simple way to show that the sea floor has always been where it is today or that, on the other hand, some kind of motion has taken place.

Several of the scientists on the leg were Europeans, a geologist from Paris and paleontologists from Stockholm and Zurich. One Chief Scientist was Dr. George E. Sutton, a geophysicist from the University of Hawaii.

When ashore, the other Chief Scientist, Dr. Joshua Tracey, works for the U. S. Geological Survey at their building in the hilly, campus-like atmosphere of the old Bureau of Standards grounds. His Pacific experience began in 1947 when he was at Bikini and scientists drilled through the coral on the atoll to reach basalt and try to prove Darwin's theory of atoll origin. Dr. Tracey also worked on Guam for the survey, living there from 1951 to 1954 with his family. He aided in mapping the island, but his own most important work was in oceanography and biology. His own particular scientific interest over the years has been the geological history of the Pacific in the past 100 million years and especially tropical corals, the island builders.

Tracey believes that the U. S. Geological Survey should involve itself in the DSDP because of phosphate, which is widely used as a fertilizer. Phosphate deposits in the sea occur along the equatorial belts of high productivity. They also occur on the west sides of continents that are areas of upwelling, cold nutrient water that rises from the deep and provides a rich feeding ground for the microscopic plant and animal life whose remains are the source of phosphate. Using the concept of continental drift, predicting where the west side of continents were in various ages, Geological Survey people have found phosphate in Turkey and in Australia. Australia had to import all of its phosphates, apparently having none of its own, so the Australian government asked the Geological Survey for suggestions. These geologists predicted that it could be found in the North Cape area of Australia. Looking into well drillings there, considerable evidence of phosphate appeared, and further work showed enough deposits to be mined for ten to twenty years. One ultimate result of the *Challenger* epic may be to find important deposits of phosphate on the deep sea floor.

Operationally, the *Challenger* on Leg VIII first drilled southwest of Hawaii to provide information on a basin in that area and thus add to the broad Pacific survey. A turbocorer, developed for the Mohole project, was also tried out on this site to see if it could solve the continuing chert problem. Unfortunately, it did not work out. The gear, whirling at 1,800 RPM in 18,000 feet of water, suddenly slowed down to zero RPM. Then a site was drilled due west of the first site on the planned north-south line to have a later reference point for sediment comparison. Two holes planned for the cruise were not drilled because the *Argo*, surveying in advance, reported that the sediments were too thin to bury and stabilize the drill string. Not only was damage likely to occur, but also experience had taught the scientists that core recovery was very poor unless the drill had penetrated at least 200 feet into the sea floor.

Once on the 140° West line, the first two sites were placed on the north and south sides of the Clipperton Fracture. This is one of the great cracks on the Pacific sea floor that dissect the East Pacific Rise. The idea was to find out the age and nature of the basement rocks opposing each other. The south side was about 2,000 feet shallower than the north but Eocene chert prevented the drill from reaching the basement. Several thousand feet of cores were taken and 15 feet of chert were recovered for minute laboratory study. Here evolution of the mysterious chert began to sound simpler. Ooze at the surface, particularly if it is full of calcium, gradually hardens as it deepens and then it grades into chalk. Farther down the chalk becomes limestone and then the limestone in even deeper sediment grades into chert. But chert is made up, in large part, of silica rather than limestone, and why, physically and chemically, the whole process occurs still does not seem as clear as the sequence appears.

All along the 140° line, the great bulk of sediments were thick chalk ooze. The last site on unlucky Leg V was also on the 140° line and it, too, found thick chalk ooze. The thickest section of this chalk ooze appeared at 4 degrees north of the Equator and thinned both to the north and south. Down to the chert, it was all about the same age. For paleontologists these cores along 140° West provide work for years to come. The sites produced a nearly complete sequence of microscopic fossils from the Eocene epoch to the present day—the stages of evolution to be read eventually like the pages of a book. There was a sufficient difference in the sequence of fossil species, however, for the researchers to consider them different formations, so they named the northern one the Clipperton Formation and the southern the Marquesas Formation, following the rule that names for such features be taken from well-known geographical features (as compared to the formation naming on Leg III). Both the Clipperton Formation and the Marquesas Formation are larger than the United States. That two different formations exist shows that in times past this part of the world ocean had two separate regimes.

Because of the chert, the drillers only reached the basalt twice, on the two southernmost holes. Over the whole track, nevertheless, a picture was produced of the composition, thickness, and rates of sediment accumulation over the last 40 million years. A considerable change in sediment accumulation happened during that time. In the early part of this period, sediment built up rapidly and then it dropped off. In general, the rate of sedimentation was two to four times faster during early times than in the past few million years. Dr. Tracey believes there has been a drastic change over much of the Pacific. Something or other caused a revolution in oceanographic conditions, in the flow of productive oceanic currents. One thought Tracey has is that it might be due to the closing of the Isthmus of Panama, closing off the Atlantic link to the Pacific and causing a great change in the Equatorial current. The trade winds could no longer push water between North and South America. The time when North and South America were joined has been dated quite differently by various authorities. Some put it as long as 60 million years. Others suggest a closing about 5 million years before the present. Perhaps the fact is that it has opened and then closed again more than one time. Dr. Tracey also suggests that the changes in the oceanic regime may have been due to some geologic events along the coastal regions, but he makes more of a point of the Isthmus of Panama's influence.

A problem that makes understanding all these sediments difficult is that the depth of carbonate compensation, the depth at which animal skeletons made of calcium carbonate will be dissolved by sea water, was more than 3,000 feet deeper some 30 or 40 million years ago than it is now. This produced much thicker sediments for that time and the belt was 25 degrees wide, from north to south. Then, with geologic time suddenness, the belt was only 5 or 6 degrees wide. Does this relate somehow to the closing of the Isthmus?

The *Challenger* sailed into Papeete, Tahiti, on December 2, 1969, after traveling 4,616 nautical miles, drilling 17 holes, recovering 178 cores out of 183 tries, and carrying with it 4,000 feet of cored sediment.

Even before the unusual-looking *Glomar Challenger* tied up at a dock in Tahiti's famous capital, the harbor had twice been witness to the DSDP. The Scripps research vessel, *Argo*, had twice been in port during her cruises, in advance of the big ship, surveying appropriate sites. Like her counterpart in the Atlantic, Lamont's *Vema*, the *Argo* was a necessary but unheralded part of the vast program. Traveling endless miles over the Pacific, the *Argo*'s voyages were not without interest. After first calling at Papeete, the ship did reflection profiling,

echo sounding, and took magnetic readings at two drilling sites northeast of Tahiti for the benefit of Leg VIII.
Then the ship stopped briefly at seldom-visited Pitcairn
Island to disembark four U. S. Air Force personnel engaged in geodetic survey work. Dr. Russell Raitt of
Scripps, a geophysicist who had "made earthquakes"
(seismological tests) all over the Pacific on a great number of cruises, was ashore at Pitcairn and, somehow—this
is surprising—suffered a broken leg. Dr. Gerald Morris
took over as Chief Scientist while the *Argo*, after the
ship's doctor had set his leg, took Russell Raitt 300 miles
distant to Mangareva Island. From there, the French
Polynesian government took over and saw that he was
airlifted to a French military hospital in Papeete. From
there, in the company of a colleague from Scripps, Dr.
Hugh Badner, Dr. Raitt was flown back to San Diego.

From Mangareva, the *Argo* sailed southwest 600 miles
to survey a possible active undersea volcano that rises
13,000 feet from the ocean floor and reaches to within
125 feet of the sea surface. This volcano had been discovered several months earlier by Dr. Rockne Johnson of the
University of Hawaii, who had noticed seismic disturbances that appeared to come from this area. Dr. Johnson
predicted the volcano's existence from seaquake records.

Rock samples were dredged from the top of the volcano and water samples taken from its top and sides,
but listening devices lowered into the water failed to detect any extraordinary sounds. "We were disappointed
not to hear anything unusual," Dr. Morris said, "but just
as glad no larger disturbances occurred while we were in
the area. No drilling sites were discovered en route back
to Papeete that met DSDP drilling requirements and the
Argo's surveys for Leg VIII ended in Papeete on November 12.

For Leg IX, the *Argo* then surveyed six drilling sites along the Equator. On this leg, from Papeete to Balboa, Canal Zone, there were three lady scientists from Scripps. From Balboa, in anticipation of the future, the *Argo* made several surveys over the East Pacific Rise off South America and then sailed to Acapulco. It was on the way from this resort to San Diego and home that a crew member, William Strickland, suddenly developed internal bleeding. The ship's doctor gave Strickland two transfusions with blood donated by crew members, and then a U. S. Coast Guard plane out of San Francisco airdropped six more pints of blood for him. Strickland improved, but the ship was diverted to Manzanillo, Mexico, where he was hospitalized briefly before flying home. The *Argo* finally reached San Diego on February 19, 1970. Not all oceanographic cruises are quite so medical as this, of course, but they tend to be colorful. The point of this story might be that it is heartening to know that men in trouble are so well taken care of, but it is also to try and give a little bit of the atmosphere of the oceanographic work that goes on, day to day, aboard hundreds of ships all over the world.

The *Argo* was out scouting ahead when the *Challenger* left Papeete on December 6, 1969, on its last planned trip, Leg IX. For the first time the cruise had only one Chief Scientist, Dr. James D. Hays of Lamont. When asked why there was no other Chief Scientist with him, he replied, "I guess no one else they asked wanted to spend Christmas at sea." He added that his wife did not care too much for his being away for two months on a voyage but if he must, she would rather he do it in winter so they could have their summer months in the Adirondacks, the mountains near where he grew up and where his family had a cabin.

Jim Hays learned nature at first hand near his home in the mountains and his questions about everything led to all sorts of reading on the subject. His family supplied him with all the books he needed and he grew up never doubting that science would be his field. He chose Harvard as a college for a good science background and decided to specialize in astronomy. But astronomy seemed too full of long calculations and short on action. One professor interested him in geology and all at once the mountains of his boyhood made sense. He not only loved them, he began to understand what made them.

For military service he chose the Naval Reserve, hoping to get to the Far East. His two years at sea, however, consisted entirely of duty aboard a radar picket ship off the coast of the state of Washington. The ship's regular course involved crossing and recrossing the Cobb Seamount, an enormous underwater mountain not far off the Washington shore that rises to within 100 feet or so of the surface. Seamounts, to some extent, were still news at the time and part of the rapidly developing field of marine geology that most universities had only begun to pay attention to. Being at sea and, incidentally, learning some seamanship, and studying the seamount he saw so often, Hays began to consider the world of oceanography. After his discharge he took his savings and made a trip around the world. Then he went to Ohio State for graduate study in geology and, since that school leads all the country in Antarctic research, his mind turned toward the Antarctic seas. He wrote his doctor's thesis on the radiolaria found in the sediments of the "Southern Ocean" and what they told of its history. He took a number of cruises aboard the *Eltanin,* a United States research ship that has done heroic work around the Antarctic for a number of years. (The *Eltanin* is funded by the NSF, as is the *Glomar Challenger*.) Later at Lamont he made many cruises aboard the *Vema*. Piston cores taken by the *Vema* in Antarctic waters were part of the evidence used in the magnetic reversal work that, in time, led to the *Challenger* cruises.

At Lamont, this new way of thinking was led by Dr. Neil Opdyke. Neil Opdyke, along with Jim Hays and two others, wrote a paper in 1966 about the study of Antarctic deep-sea cores. The very revealing subtitle was "Paleomagnetic study of sediments in a revolutionary method of dating events in Earth's history." Having been in on the Project at the very beginning, as it were, and given his experience, it must have seemed only reasonable to name Dr. Hays as Chief Scientist on a *Challenger* cruise.

In a way, the drilling was uneventful. They reached the basement at eight out of nine holes. They found no chert because the basement was comparatively young. The only adventure, if it really was one, was the occasion when it seemed necessary to drill to the basement, rather than core on the way down, because they were in a strong equatorial current, the bow thruster was broken, and the ship could not be held in position for long. Hays decided to make the best of it. With a tight schedule, he did not want to miss any of the next positions. The bow thruster remained broken the rest of the leg, but fortunately the currents, always uncertain, were no particular problem because they were weak from then on until Balboa.

The first stop on the leg had been north of the Tuamotu Islands to sample the floor and date the archipelago. This had nothing to do with the major objective but, the ship being in the vicinity, the advisory panel decided to seize the opportunity of learning more about a portion of the Pacific they might never have the chance to drill deeply into again. The drill was stopped by turbidites after only 90 feet where the sediments were lower Eocene, around 50 million years, so the archipelago is at least this old. Some day, when a scientist wants to present a hypothesis about the formation of Pacific island chains, he will have this particular fact to reckon with.

The major idea of Leg IX was to make a track along the Equator, but sampling to the north at times, running from west to east. This is the present zone of high productivity and, therefore, rapid sedimentation. One jog was made to 7° North, a bit above the Clipperton Fracture, and comparison of basement ages there with known ages to the south of the fracture indicated that there had been a displacement between the two sides of nearly 200 miles. The sediments at the basement here presented a problem because they were baked, indicating that the lava had intruded after they had been laid down. There might possibly be more sediments buried beneath the supposed basement. To make sure this was not so, Hays had the ship steam directly south for several hundred miles to drill again. He reasoned that, if the plate had been moving directly north when the first sediments were laid down, these sediments closest to the basement should be of the same age at a due south site on the same Pacific plate. The age of the basement at this new site was nearly identical with that of the site north of the fracture. Hays felt reasonably sure he had reached the true basement in both cases.

The basic finding of the track along the Equator is that sediments to the north of it are always thicker than those to the south. These thicker, northern sediments are now beyond the equatorial zone of high productivity, and the reasoning is that this northern region must have been the Equator itself. Therefore, the plate is moving north.

Furthermore, by determining the sediment ages and measuring the distance between the drilling sites, it was determined that for a period of 25 million years in the past, the rate of sea-floor spreading had increased markedly. It became a faster and faster rate until about 10 million years ago, when the Pacific spreading rate was about four times that of the Atlantic. Then, abruptly, it slowed from an increase of 5½ inches a year to about 2½ inches a year.

To explain this, Dr. Hays drew upon a theory proposed a few years before in which it was suggested that sea-floor spreading was caused by gravity, which pulled the ocean plates into the deep trenches at the ocean's edge. Gravity may have pulled the Pacific floor plate down at the trenches with increasing speed until a critical point was reached approximately 10 million years ago. For some reason, the plate could go no further. "At that time," Dr. Hays said, "the entire moving plate apparently shifted direction."

Evidence of this shift was found in the history of sedimentation rates of the cores taken on or near the equatorial line. Some authorities have extrapolated this and say that what was the Equator in Eocene times is now at 40° North, or about 120 miles north of San Francisco. Plotting the rich sediments showed that they had been carried slowly westward and then, 10 million years ago, the direction changed to northwest.

In explaining his idea of sea-floor spreading, Dr. Hays said, "This observed acceleration suggests the possibility that the principal Pacific ocean plate was drawn by the force of gravity pulling the edge of the plate down into the earth at the deep-sea trenches near the Asian continent." He made an analogy. "It's like laying a terry cloth towel on water. It sinks first at the edge, and the more the towel goes under water, the faster it sinks, and the faster the part on the surface is pulled along." It is known that the sea floor does turn down at an angle in the huge trench systems. Hays believes that the plate, which runs from the Aleutians to New Zealand, goes down into the drain for at least 500 miles. The Pacific is under tension, not compression. Some water is carried down with the plate and this is important in mobilizing volcanoes. In support of his figures on the speed of spreading in the Pacific over the last 10 million years, 2½ inches per year, Hays points out that this is the rate of motion of the San Andreas Fault.

As has been mentioned, a popular theory explaining sea-floor spreading is that it is moving like a conveyor belt with convection currents of hot rock moving just beneath the earth's crust and carrying the sea floor on its back. Dr. Hays believes that the Leg IX findings do not support this idea very well. To make the Pacific floor move at the varying speeds found on this cruise, the currents would have had to move in a stop-and-go fashion. Dr. Hays finds it difficult to explain how great masses of boiling rock could speed up and then slow down. In addition, he points out that the great plates of the world are moving in contradictory directions. They have no pattern, as would be likely with great convection currents; in fact, each plate seems to have had a quite independent motion.

The thought has also been expressed often that basalt lava wells up from the earth's molten interior and is extruded along the great ocean ridges, pushing the sea floor apart as new rock is added to the edges of the plates. Many studies in various parts of the world ocean seem to show that this process has been fairly constant.

Yet the evidence from Leg IX shows that the Pacific motion has not been constant and that this one ocean floor plate, at least, may have been pulled rather than pushed. Dr. Hays believes that more research may show all three kinds of mechanisms at work; the pushing of upwelling rock, the carrying of convection currents, and the pulling of gravity, at various times and in various combinations through the geological ages.

In the words of an English geologist, Dr. David Elliot, soon after he had participated in findings perhaps as sensational as those of Leg IX, "One has moved the unknown one step farther back." Before the *Glomar Challenger* reached Balboa in the Canal Zone, at what was originally planned to be the end of a very long journey, the DSDP had received a three year extension of the deep ocean probe. So much had been learned that it was realized once more how little anyone actually knows about our earth.

14

"It isn't easy to get money for a project like this. This project won't kill anyone!"

—JOSEPH A. CLARKE,
Captain, *Glomar Challenger*

WOODS HOLE, Lamont, Miami, the University of Washington, and Scripps all had the news that the *Challenger* could go on beyond Leg IX about the time the ship sailed for Tahiti. The NSF granted their joint venture an additional $22.2 million contract for another 36 months of drilling. Miraculously, the DSDP had kept within its initial budget of $12.6 million and had overspent no more than $100 thousand. Considering that this sum covered an 18-month period, trying something that had never been done before, with all the accidents that might happen at sea, with inflation, with so many different people involved, this achievement of keeping so completely close to their budget stunned the bureaucrats in Washington.

In announcing the new grant Dr. William D. McElroy, Director of the National Science Foundation, said that the DSDP "has proven to be an outstanding scientific and technical success. It has excited not only the more than 300 scientists of this country who have been involved in the project's planning and execution but many others throughout the United States and abroad who are following the progress of this national program. Undoubtedly, in the years to come research on the core materials will contribute to the interest and enthusiasm of graduate students, bringing many more into this field of work.

"The success of the Deep Sea Drilling Project to date has put the United States in the forefront of deep ocean exploration. The continuance of this project will contribute significantly to the world's scientific knowledge." He added that industry should greatly benefit from the project because of the technological advances that had been made and because of the information on natural resources (the discovery of oil in the very deep ocean and new ideas on how valuable metal ores are formed) that had been produced.

Dr. William A. Nierenberg, Director of Scripps, added that everyone was naturally gratified but "there is much more to be done both geographically and technologically."

Everyone was happy. Kenneth E. Brunot, the Project Manager and the man most burdened with the technical problems, was quoted, "Like any new project, we have encountered complex problems but our project people come from industrial, scientific, and governmental circles and they have worked beautifully together to solve them." Global Marine Inc. came in for its share of praise.

Those thin layers of chert remained as an obstacle, but Brunot remarked that with the new grant they might go ahead working on a system of reentering a hole after changing a worn-out bit and get the better of this annoying obstacle.

Terry Edgar, speaking for the scientists on the DSDP, spoke again about one of the original concepts of the idea. "It is difficult to imagine another project of national scope that might produce such profound results for so many branches of science."

As for records set, remembering that operational offshore oil wells at present seldom work in more than 600 feet of water, the *Challenger* had kept station in depths up to 20,140 feet of water in good and bad weather. It had let out as much as 20,742 feet of drill string and the drill had penetrated into the ocean floor as much as 3,231 feet. Working at sea, there had been scientists from fifty-nine United States universities and research institutions, five from the petroleum industry, four from the government, and nine from foreign research institutions.

After all the congratulations, it was time to begin planning ahead. In the largest way of thinking, the ship, having through its work established the idea that the sea floors are spreading and that continents move about, would now stick closer to the shores, examine the remaining large ocean and the smaller seas, and start to fill in the details. It would cross the Atlantic again, make a tour of the Mediterranean, return to the Americas and take another look at the Caribbean, probe the northern Pacific, the South Pacific complex east of Australia, the East Indies, the Indian Ocean, the South African regions, and finally return to Galveston, Texas. At that point, as mentioned in the scientific trade magazine, *Geotimes*, ". . . we may expect a lot of new knowledge and probably some new problems concerning the structure and history of the Earth."

Ungraciously, even as *Geotimes* noted the extension of the project, it also printed a letter from D. B. Stone of the Geophysical Institute, University of Alaska. "Noting that the drilling of Challenger Knoll on one of the JOIDES cruises produced an oil-saturated core, one wonders whether the project has any contingency plans in the event that future drilling starts an oil seep? . . . the starting of an unstoppable oil leak could well have a catastrophic effect on life on this planet."

Indeed, the Project did have plans, and even if they had none, the federal government, which, after all, was supplying the money, was certainly aware of the problem. The beaches of Santa Barbara were still covered with oil and an off-shore well of the Chevron Oil Company off the coast of Louisiana was on fire as the *Challenger* entered the Gulf of Mexico for Leg X. In addition, executives of the NSF were testifying before congressional committees for their annual appropriation at exactly the same time. D. B. Stone of Alaska need not have worried. Quite a number of people, besides those on the ship, were watching the *Challenger* (a child of the NSF) as it drilled holes in the well-proven, oil-rich ocean floor of the Gulf of Mexico.

Dr. Worzel, who had been Co-Chief Scientist for the Leg I cruise that found oil in very deep water on the Sigsbee Knolls, had the same role on Leg X. Dr. William R. Bryant of Texas A & M, but also representing the Gulf Universities Research Corporation (GURC), was the other Chief Scientist. A number of oil companies sent knowledgeable men to do the work in this region that was of considerable interest to each of them. As for objectives, Dr. Worzel said, "We would expect to settle geological debates on whether the Sigsbee Basin is a foundered piece of continent or has long been an oceanic area, the nature of the bounding escarpments, whether a salt layer underlies the whole basin—and something about the history, extent, and origins of the salt dome province, with its associated petroleum and sulphur accumulations, discovered on Leg I."

The worst conditions of the *Challenger's* career so far, according to Captain Clarke, came when they tried drilling while being roughed up by the Gulf Stream in the Florida Straits. The current was coming at them at speeds of 4 and 5 knots while they tried to stay in position to get a core. To make matters worse, the acoustic beacon at the bottom, on which they relied to know their position over the hole, began to fail them. Steering manually, they did manage to obtain a core in spite of the Gulf Stream.

A food problem occurred on the voyage when the cooks discovered they had been sold bad meat in Galveston. For nine days they had to serve variations on frankfurters and sausages. They were, as always, inventive and inspired, and there were enough steaks and prime ribs to give the crew one serving of beef every day, hardly starvation rations but for the *Glomar Challenger* a definite debasing of culinary standards.

Fifteen holes were drilled at 13 sites, 3,000 feet of cores were brought up for evaluation, and at almost every site they ran across methane gas. Every hole on the Leg was cemented except where the drill pipe made this impossible. The operations manager kept a very firm hand. Probably the great water pressure would have prevented a blowout or stopped any seepage but no risks were being taken. The DSDP had appointed a special advisory panel for this leg who considered it risky to drill into sediments on seismic features, ones that showed any signs of earthquake activity, or to penetrate into several thousand feet of sediment without being able to reenter, if more cementing should be necessary. The panel recommended that some test holes be relocated and that others be postponed until some other time. The NSF agreed to the recommendations on seismic features but did permit deep drilling on the abyssal plain. It did, also, accept the Department of Interior's new, stringent rules for drilling into potential oil traps, knowing that within a year there would be the reentry capability needed to make such operations safe.

According to some reports, the situation aboard the *Challenger* in the Gulf of Mexico was not very pleasant. Messages apparently were even radioed ashore making demands and counterdemands. The scientists on the ship wanted results but, rather than have the whole DSDP shattered by some highly publicized incident of pollution (the United States public being acutely aware of oil pollution from the sea just at this time), other authorities evidently decided that it would be wiser to accept a bit less in the way of cores. The success or failure of Leg X was not so important that it should be allowed to destroy the entire future of the *Glomar Challenger*. One angry scientist is said to have held a press conference when the ship docked in Miami in which he said that one result of the safety requirements was to "emasculate a good part of the program that was going on in the Gulf."

Nevertheless, a rather pithy press release was issued for publication on landing. In part, it read, "We found nothing in the Gulf to support or deny the theory of continental drift," said Dr. Worzel. "Our findings do deny any continental drift occurring there for about the past 100–135 million years. Beyond that I cannot say, but up to that time the deep basin was there and existed in essentially that same environment and location as it does today." Dr. Worzel is a conservative scientist who is not yet convinced of the new theories.

The cores recovered will provide a continuous dating for a mix of two types of fossils, silica base fossils mixed with limestone type fossils, from about 30 million to

100 million years ago. Dr. Worzel stated: "This is an invaluable yardstick for the future of correlating dates between these two types of fossils."

He also pointed out that two geologically exciting scientific findings turned up on the voyage. He said that thick coarse sand was found in the deep Gulf Basin about 25 million years ago. "These are deposits of a turbidity current and are very significant." He explained that a turbidity current is like a river of sediment on the ocean bottom. Coarse sands, or turbidites, are caught and mixed in a strong current and this sediment, which normally would be deposited in shallow water, can be carried to a deep water environment.

"Some scientists have said," he continued, "that this condition is peculiar to the Pleistocene Ice Ages, but there has been some evidence from corings elsewhere that this is not so. Now, here in the Gulf, for the first time, we found the coarsest turbidites I have seen and evidence of very strong turbidity currents.

"Geologists will have to explain why turbidity currents were stronger in the Miocene, a period that had no ice ages."

The press release went on to say, "The expedition also found a history of varying chert deposits which Dr. Worzel said is very important for geologists seeking to explain how and why cherts are formed. . . . This has been a puzzle to geologists, who up to now have not encountered enough chert gradation in an area to allow full study." Dr. Worzel also mentioned that the natural gas, mostly methane, was found at all holes drilled in the deep basin. He said it was apparently biological in origin and seemed to be strongest at about 300 feet below the bottom. He thought this should interest all geologists seeking to understand what causes oil to form.

Interviewed by reporters in Miami, Dr. Helen Foreman, a paleontologist from Oberlin College in Ohio, said, "I enjoyed the operation very much."

The *Miami Herald* of April 6, 1970, carried a story about the *Challenger* as it lay alongside a pier at Dodge Island, commenting on its first visit to Florida. "Already this 400 foot ship has built up a solid claim to being the most important research vessel in the world. And it's still drilling. The entire project makes up the most ambitious earth sciences research operation the world has ever seen."

During Open House Day when the University of Miami acted as host, Dr. Richard Bader, one of the Chief Scientists on Leg IV, called the ship "a mighty tool. In the bottom of the oceans resides the debris of the eroding land, a history of the earth, its life and contents. This ship has recovered vast stores of sediments and vast stores of knowledge that will take many, many years to fathom and understand."

Mel Peterson, who flew in from San Diego for the occasion, spoke of the fact that America and Europe broke away about 200 million years ago. "One of the things we want to find out is *just when* this did happen. Ocean currents near the bottom have stripped away much of the usual sediment layer and have given scientists a better chance to power the diamond drill bits into the primordial sea floor.

"That's what makes this area [Florida] so juicy for us. Nature has removed most of this stuff by itself. In just a few thousand feet, the Atlantic's very floor could be penetrated."

"Hundreds of representives of South Florida oil companies, ocean industries, scientific organizations, friends of the School, and students of oceanography were given tours by enthusiastic guides who flew in from Scripps and Global Marine Inc. for the event." This was the report of the little monthly bulletin, *26° N 80° W*, put out by the Rosenstiel School. When the courtesy call was over, the *Challenger* went down the Miami Ship Channel and headed for New York.

Leg XI turned out to be a quite prosperous expedition. A major objective was to find out just what happened to the crust of the earth as the continents tore apart. The thought was to examine this event through investigation of the oldest sediments in the Atlantic, those formed just after the separation. Since Leg I, which covered roughly the same area, a chink had been discovered in the armor of the chert that had been such an obstacle to drilling. Seismic surveys over the region by Lamont had found places, which were named windows, where the chert was missing and through which the basement might be reached. Improved equipment and added experience also added to the hopes for better performance.

One Chief Scientist was John Ewing, Dr. Maurice Ewing's younger brother, who studied at Harvard and Woods Hole before joining Lamont.

His Co-Chief, Charles D. Hollister, studied geology at Oregon State and received his doctorate at Columbia University in 1967.

The one lady aboard for this trip was Miss Paula Worstell, a foram paleontologist from the DSDP in San Diego. France was represented by Dr. Yves Lancelot, sedimentologist from the University of Paris. Dr. Lancelot comes from a village in Brittany that is actually called the Knights of the Round Table. In addition to the Lancelots in the neighborhood, there are two other families bearing last names that are the same as those of King Arthur's knights. Dr. Yves' brother once traced the family back to the fourteenth century but could find no records older than that. The oceanographer will probably never discover whether or not he is descended from Sir Lancelot but he feels that it is so.

The drilling began with excellent weather in the northeast Providence Channel in the Bahamas. The discovery of very ancient reef sediments, covered by deposits obviously laid down in an open sea, indicated that the Bahama block has been gradually subsiding but that an upward growth of nearby reefs has kept pace. Moving closer to the site where the oldest sediments so far had been found during Leg I, the men on Leg X went down into the basalt and found Jurassic limestone right above it. This they dated at 170 million years and, for the location, felt this was the oldest possible date. The old objection might be made that older sediments still exist beneath the basalt but the scientists feel their profiles show nothing farther down than more layers of basalt. The existence of limestone shows that the deepest sediments were laid down in shallow water, in a very young ocean before it got very deep. An almost exactly similar hole was drilled a few hundred miles southeast of New York but the paleontologists thought that it was slightly younger. Asked at a press conference what age the basalt showed by radioactive dating, John Ewing and Dr. Hollister explained that the radioactive technique will not work on basalt if it has been weathered, as this had. A new method is being developed, however, by which the ancient glass, often present along with basalt, may be measured.

At the site off Norfolk, continuous coring through 1,000 feet of sediments down to upper Jurassic produced an extremely rich find of nannoplankton, ocean floor foraminifera, radiolarians, dinoflagellates, ostracods, polled, and spores. These presented an unparalleled opportunity to study these fossil groups.

This same site produced, at a level perhaps 50 million years younger, a spectacular sequence in mineraliferous and metaliferous clay beds. These brilliant red, yellow, brown, olive, and orange colored layers contained a variety of minerals including zeolites, micas, quartz, siderite, pyrite, rhodocrosite, sphalerite, amphiboles, and iron-manganese nodules. To informed observers, these looked very much like the famous hot brine deposits found a few years ago in the Red Sea. There were very high salinities far up the sedimentary column. A thin veinlet of native copper was also found about 45 feet above the basalt. The finding of all these metals is a matter of great excitement to geologists. One of their great questions has always been how metallic ores concentrate. Why Butte, Montana, for instance, is one great hill of copper. And this complex display from Site 105 will, perhaps, show the role of the sea in the process. Near the basalt, there was considerable iron, probably formed when the sea floor was hot.

At three of the sites there was a disconformity, an absence of sediment for a period of 70 million years. Presumably some kind of large-scale erosion removed this 70 million year record of deep ocean history. At several locations a jet black clay was found, called carbonaceous, that might be "pre-oil" but not "pre-coal." Tests showed that it would burn. This, naturally, will all be studied with a great deal of care.

Some of the drilling was done on a "mountain of mud" that stretches up and down the East Coast of the United States. Although it is several hundred miles of sea, Dr. Hollister described it as "Himalayan in size." Off the Bahamas it is 6,000 feet thick, but off the coast of New

York it is 27,000 feet thick. No one, as yet, knows how far north this great range extends, but it is a far larger mountain system than the Appalachians. It is the thickest sediment in the western Atlantic and how did it get there? Presumably it is derived from erosion long after the continents separated, probably Miocene or around 20 million years old, and it must have derived from the north. Dr. Daniel Habib of Queens College, aboard the *Challenger* as a specialist in ancient seeds, a palynologist, says that the pollen in the southernmost cores resembles a modern spruce tree but spruces never lived in Florida during any primeval period. The French geologist, Dr. Yves Lancelot, said that the cores show strong similarity to some in the Mediterranean and particularly to rock outcrops in the Apennine Mountains of Italy and Sicily.

All during the trip the weather acted beautifully—except at the last site. Here a low pressure area developed, a storm moved in at a direction abnormal for Atlantic low pressure areas, and the sea swells grew 10 to 12 feet high. Working conditions became impossible with the ship rolling 12 degrees and pitching 5 degrees. While pulling out of the hole to clear the ocean floor, the *Challenger* was forced approximately 1,500 feet off location by the strong Gulf Stream current with six stands of pipe remaining below the mud line.

But, as the Operations Resume states, "Due to excellent crew performance and cooperation of all personnel concerned, new records were established to add to the growing success of the Deep Sea Drilling Project. The oldest sediment so far was recovered, a new record for penetration (3,332 feet) was made, and the most basalt recovered (42.7 feet)." The turbocorer, developed for the Mohole, was used with some success. Tungsten carbide roller bits were used, rather than diamonds, and these proved to be invaluable.

A stopover of 11 hours and 44 minutes for supplies and a change of crew and technical team was made in Norfolk, Virginia, on May 11, 1970. Then the *Challenger* sailed to its Hoboken mooring in New York harbor. There it would be outfitted with new gear to try out a new and much needed capacity.

15

"Something new hardly ever works the first time."

—ARCHIE MCLERRAN,
National Science Foundation

"Re-entry will not be considered."

—CONGRESSMAN JOE EVANS, TENNESSEE
(Strong opponent of the Mohole)

"This reminds me of a blind elephant trying to take a drink out of a teacup."

—DR. MELVIN N. A. PETERSON,
Chief Scientist of the Deep Sea
Drilling Project

"Did you ever try to drill a well in more than 15,000 feet of water. What if you had to *make a trip* to change the bit? Think you could ever find the hole again? It's *never* been done before but someone's going to try!"

—*Oil and Gas Journal,* June 15, 1970

WHEN VIEWED at dawn, from another ship far out at sea, the *Glomar Challenger* presents a very imposing picture. The hull is black, the derrick very high, and the impression is one of bigness. From a mile or two away, alone with nothing else in sight but the sea, the sky, and the horizon, the ship looms and has an air of authority about it. Along with representatives of the Phillips, Esso, and Standard Oil companies, and a man from International Nickel, this writer boarded the *Challenger* 180 miles southeast of New York while the reentry trials were underway. Ever since the chert, unsuspected in what geophysicists thought was the uniformly soft deep ocean floor, had begun to frustrate scientists, the ability to get back into the hole after a bit had been worn out seemed a terrible necessity. If the drill could never get through the chert, how would it ever be possible to reach the true basement? If, on reaching basalt, the drill could not penetrate more than a few feet, how would it be possible to prove to skeptics that the rock in question was not just a later addition, a subsequent flow of lava, and that older sediments were not farther down and simply covered up?

A number of ideas on how to find a hole only 10 inches in diameter, at the bottom of 20,000 feet of water black as night, had been considered during the Mohole planning. When reentry became a matter of urgency soon after the *Challenger* began drilling, the experts went to work and here at last was the equipment ready to be tested, devices based largely on the ideas for Mohole. The ship had taken this aboard at Hoboken and almost two weeks were allotted to try it out at a position that was chosen because it had at least 10,000 feet of water, was not in the Gulf Stream, was near both Boston and New York, and the bottom sediments resembled deep ocean basins but had no chert.

The group boarding the ship out at sea on the test site had been brought there by the U.S. Navy ship *Fort Mandan*. The agreeable captain of this ship boasted that he had sent a liberty party of more than 100 men ashore in Jamaica and they had all returned without any of them getting into a brush with either the Shore Patrol or the local police. (Navy men will admire this record.) The *Fort Mandan* sent the passengers over in a landing craft. Many of the ship's sailors had wanted to take a look at the unusual vessel, but the seas were running too high to transport large parties so the men had to be disappointed. The visitors boarded the *Challenger* by a ladder over the ship's side that could only be caught when the landing craft rose high on an upswell.

The visitors were given a brief tour of the *Challenger* by Darrell Sims, Project Engineer, then invited to the science lounge to listen in on a conference having to do with the failures of some of the trials. The worst problem was that the reentry cone, which really looks like a funnel, had broken loose while being fitted onto the drill pipe and had plunged 10,000 unrecoverable feet into the sea bottom. There was only one other cone on the ship.

Mel Peterson and Terry Edgar from Leg II were aboard for these trials and had hoped to get back to Site 8, which had been so troublesome, and this time make a penetration with the reentry gear. It was already clear, however, that Site 8 would not be revisited on this trip. The conference was on how to strengthen the second and last cone so that it, too, would not be lost. In addition to reinforcing the cone, elaborately detailed plans were made for every step, every movement, every line that would be used while the cone was keel-hauled over the side and down under the moon pool, where the whole drill string assembly is lowered through the ship and where the reentry cone has to be attached to it. After a very short, agreeable, businesslike discussion, plans for action were made and all hands retired to lunch.

Almost the entire DSDP permanent staff had abandoned Scripps at La Jolla to be on board the *Challenger* for the reentry tests. In addition to Mel Peterson, Terry Edgar, and Darrell Sims there were Ken Brunot, Project Manager; Swede (V. F.) Larson, Operations Manager; Archie McLerran; Tom Wiley, Public Information Officer; and Mrs. Sue Thompson, actually Edgar's secretary but signed on as "Yeoman." Pretty, young Sue Thompson had a baby not more than a year old, but she left him in her husband's and mother's care for two weeks. She had never been to sea before. Also from Scripps were Larry Lauve, photographer, and Fran L. Parker, paleontologist, a displaced New Englander who after twenty years in California still had the air and sound of educated Boston about her. Dr. Nierenberg, head of the whole Project, remained in La Jolla but he was in frequent contact by radio.

Basically, the reentry scheme has three elements. One is the reentry cone, in the shape of a funnel. Sixteen feet across at the top, it tapers down to the hole at the bottom, which is just wide enough to admit the drill string assembly. This cone is fitted onto the drill string while it is just underneath the ship. Then it is carried down and rests on the sea floor while the drill string, with its drilling bit and plastic liner to catch the cores, penetrates deep into the sediments below. When it is considered time for the drilling bit to be replaced, after it has been worn down by drilling in chert or basalt, the drill string is pulled up, leaving the reentry cone behind when the cone and string are disengaged from each other by an acoustic signal from the surface. After the drill string has been brought back to the ship and a new bit fitted on, the drill string is lowered again, this time carrying a sonar sending and receiving device to locate the cone. During the tests there were three reflectors on the rim of the cone that would send back signals to the sonar on the end of the drill pipe to indicate its position. All of this signalling could be seen on sonarscopes on the ship's bridge and at the drilling platform. This scope looks very much like an ordinary radar screen. From the bridge, the captain could see on the scope when the end of the drill string was in relation to the cone and maneuver his ship to get the two as close together as possible. For the last maneuver, there was a hydraulic jet at the end of the drill string, which could be manipulated from the drilling platform, that moved the drill string right over the hole. At that moment the drill operator quickly lowered the whole drill string into the cone. The shape of the cone guided the drill string right back into the hole that had previously been drilled. The sonar reflecting system worked on a radius of 500 feet. The whole bit changing operation should take about 15 hours.

At least, that was how it was supposed to work. What happened was that the hydraulic system did not function at all. The acoustic system that was supposed to separate the drill string from the reentry cone did not function either and so a simple locking device was rigged by which the two could be disengaged simply by twisting the drill string from the surface. The sonar system, usually called the EDO after the name of the manufacturer who specially devised it for this work, also collapsed when subjected to the pressure of deep water.

There was no spare sonar aboard the ship and the EDO representative, who had come to observe the tests of his company's equipment, thought that the situation was hopeless. He did not reckon with Project personnel. They opened up the long, terribly complicated electronic gadget and went doggedly to work. As a result of the accident, a special petroleum-like fluid that had been part of the works escaped and there was no more of this kind of fluid available. Darrell Sims, who had spent twenty years in the oil business, smelled every sort of oil that the ship carried for a variety of purposes and tested each of them for conductivity. At last he found one he liked and cut it with paint thinner.

Six men, including both Chief Scientists, worked round the clock, long agonizing hours during which everyone simply had to wait around. At last they got the absolutely essential sonar working again. Terry Edgar remarked at this point, "I've been on many oceanographic ships but I've never seen people tackle a job like this or come up with an answer. All the men contributed greatly. No *one* could have done it alone."

With the EDO now repaired, a crane hoisted the 16-foot-wide reentry cone into the air and with two men riding on it, over the side and set it gently in the water. It was firmly secured with all sorts of lines. When the first cone had been fitted on, everyone had been preoccupied with how it would work on reentry and it was perhaps this preoccupation that caused the accident to the first cone. This time there were no mistakes. Every move had been carefully planned and gone over in advance. Each roughneck knew exactly what he had to do and the execution was practically flawless. There was one small hitch toward the end of the maneuver, but the most eager of all the hard-working roughnecks hurried down into the moon pool. He was rather slight but, as Ken Brunot said, "strong as a bull," and he soon set matters to rights.

With the cone in place, 240 feet of casing pipe began to be assembled on the drilling platform. As the top section was being emplaced, the huge, winch-operated tongs

that give a powerful twist to tighten the sections together twisted too hard and badly dented the pipe. Re-
placing it took considerable time.

Working on something entirely new, like these tests
for reentry, requires an ability to wait, endless patience,
and in spite of all the frustrations and things that went
wrong, everybody on the *Challenger* kept his temper and
good humor. The damaged section of pipe was not the
last of the trouble.

It had become obvious several days before that extra
time would be needed to make this trip a success. A radio
message was sent to Dr. Nierenberg in California asking
for a 24-hour extension. The affirmative reply now came
in and in a short time there appeared on the blackboard
in the mess hall this news, "Jody extends *Challenger* for
24 hours." Jody stood for JOIDES, the name of the five
sponsoring oceangraphic institutions.

When the damaged section had been replaced, the drill
string began to be lowered at the rate of about one foot
per second. The drilling signal, a large red ball over a
large white diamond topping another red ball, was
hoisted to inform any ships that might be nearby. The
water depth here, according to the Position Depth Recorder was 10,014 feet.

As the lowering continued, section after section of pipe
was removed from the automatic pipe rack, and lines
from the derrick carried them over to the drilling platform where they were fitted on. This rhythmic performance, when the drillers' art takes on the form of a stylized
dance and when one crew vies with another to see who
is faster, somehow, this time, went wrong. A flying section of pipe whacked one of the roughnecks on his hard
hat and he fell to the deck. Everything stopped while
the driller in charge went over to him, but the man who

was struck insisted he was all right. He did sit down for a few minutes to recover his senses, then went right back to work. The other men called him "Superwop," a young man with carefully combed long hair who always wore rather elegant resort shirts and trousers in the mess hall, but they did not mean it unkindly.

Then, when 57 feet of pipe were out, the EDO, which had tested perfectly while on deck, at this depth suddenly conked out. It had to be brought back again to the ship and worked over once more.

Then, finally, at 3:00 A.M. everything seemed to be ready. The reentry cone had been placed on the bottom and the EDO was operating successfully on deck. By 6:00 A.M. the tool was on the bottom, but it took two hours to find the reentry cone. Captain Clarke was conning the ship and the spacious bridge was full of interested spectators. The seaman on watch brought up two big thermos bottles of coffee and numerous paper cups. At 8:30 A.M. the end of the drill string had drifted over the reentry cone several times but never stayed long enough to drop the string in. Without the hydraulic works at the end of the drill string, all the maneuvering had to be done by the ship itself, with considerable uncertainty about the reaction time—how long would it take for the motion of the ship to move the bottom of the drill string 10,000 feet below. At least the weather was good and the computer had become used to compensating for a one-knot southeast current.

The ship was being controlled by the computer, known as "Elmer." The captain would modify the dynamic positioning control by pressing buttons. Pushing the computer buttons, however, only moved the ship in increments of 100 feet: thus each new direction tended to overcompensate. The captain and all his visitors kept looking at the sonarscope. The drill string would be in perfect position when the three echoes, from the reflectors on the reentry cone on the ocean floor, made a triangle right in the middle of the scope. Several people took turns plotting the ship's various courses on a maneuvering board. After a number of positions were marked down, Mel Peterson noticed that the ship was making a figure eight over the target.

At lunch time the captain, who usually ate in the mess hall with everyone else, had his prime ribs of beef served to him on the bridge. Although he had been in personal control of the *Challenger* for hours, he insisted he was not tired. "We have until (9:00 A.M.) tomorrow to go-go-go with all systems." The object of all the maneuvering was to set up a collison course with the target, but the bearing kept changing and since the EDO did not give bearings, the whole procedure was very difficult.

By 1:35 P.M., the sea had reached state Five. This meant that things were getting a bit rough. The wind blew at 30 knots, yet it was a bright and sunny day. At 1:40 P.M., the captain ordered "Stand by" and then told the drillers to start lowering the pipe. But the time lag was too long, if only a matter of seconds. Trying to stab the target, the drill string had only hit mud.

Everyone immediately involved was tense and tired but they still managed to make a few jokes. The observers talked intermittently about the problem and then about trivial, or at least unrelated, matters. The only reality seemed to be in the flickering sonar scope itself. It was actually all quite abstract, with no one actually visualizing the objects themselves, the end of the drill string and the cone, two miles beneath the ship in the cold and almost lifeless water. The problem, in one way, seemed so little. In all the vast ocean, and two miles down, the question was only a matter of a few feet but how elusive they were.

The long afternoon wore on. A technician who had served aboard the nuclear sub, *Nautilus,* as a sonar man, suggested putting a phony depth into the computer, 11,000 feet rather than 10,000, in order to fool Elmer, the computer, and reduce the size of the possible changes to less than 100 feet. If the ship were to be operated manually, the target could have been hit long before but, with dynamic positioning, which would be necessary if a hole were to be drilled as a routine operation, the computer had to learn where the cone was in order to keep over it. Reentering the way it was tried all that Sunday was just about the hardest possible way of doing it.

The time for dinner came and then Ken Brunot, Project Manager, joined a bridge game in the science lounge. Then suddenly the man from Phillips Petroleum rushed into the room and said, "He thinks he's got her stabbed!" The cards were dropped on the table and everyone ran up to the bridge. The official time was 7:54 P.M. The crew had lowered enough length of pipe to hold firmly.

At 8:00 P.M., the EDO instrument was out of the hole and had begun its hour and a half trip up the pipe to the ship. The total weight of the drill string out was 290,000 pounds, a figure the drill operator could see on his instruments. At the moment of stabbing he could feel nothing, but he could see the impact by the instant loss of 10,000 pounds of weight. Then the weight loss was 30,000 pounds after the ship surged up and down in the sea swells, breaking the shear pins that held the drill string to the cone.

At 9:30 P.M., coring began. At 1:00 A.M., quite a number of people were out on the drilling platform, wearing their required hard hats. Larry Lauve was on hand with cameras. The flood lights brightened the dark night and gave a movielike atmosphere to the whole scene. The core itself was about to appear. Then suddenly it rose through the hole in the drilling platform, all dripping and cold from its long trip up from the deep. Charles Simons, the drill operator, left his controls, went over to the prize, took off his hat and hailed it "the first core from a reentered hole. Part of a world-famous operation!"

Pictures were taken and everyone went about grinning. The core was extruded from the plastic liner by means of an air pressure pump and, as it came out, the technicians caught it in five-foot long rounded plastic trays and carted it off to the laboratory. Peterson and Edgar smelled a sample and said it seemed to have methane gas in it. It was gray clay, probably Pleistocene (or contemporary with the Ice Ages), a conglomerate kind of sediment, lots of plankton mixed with turbidites. It had probably been bulldozed off eastern North America by the ice sheet. Ten years ago, this kind of core would have been considered a find. Now this little sample had been taken only to prove that reentry had been achieved. For the first time ever a carefully guarded *Challenger* core would be cut up, put in little plastic containers, and given to everyone aboard as souvenirs.

When the excitement died down a bit, Peterson walked back to the drilling platform and looked at the pipe that carried the reentry instrument. He commented that there should be an oceanographic museum but that most of the original instruments were either lost at sea or worn out. The science has already reached the point in its development when people can look backward to the good old days.

The *Challenger* then headed for Boston, a 41-hour trip. Now relaxed a bit, two of the principals involved found a bit of time to talk. Mel Peterson felt that, in spite of losing a reentry cone, they had learned a great deal on this trip. They had, after all, managed a reentry and this was the major problem. Modifications would now be made of the existing equipment and the revised tools, taking advantage of everything that had been learned during the tests. The changes would be made before the *Challenger* sailed on Leg XV to the Caribbean, where they could confidently expect to find chert.

Reentry being accomplished, Peterson now had a more profound worry. "Geophysics is now in danger from its own success." The science had grown so rapidly in the past ten years because it knew no boundaries; it was worldwide. Scripps, Woods Hole, and Lamont ships ranged everywhere. They developed a global view. Land geology had developed artificial boundaries. Because of the organization of the work, state universities and such, geological formations tended to start at state borders and also end there. Land geology had become small and parochial.

Now geophysics with its big view and big discoveries has attracted the attention of governments. Interested nations now want to set boundaries on the exploration of the seas. On the new Atlantic leg coming up for the *Challenger* they had wanted to drill quite close to the Canadian shore, but they had been diplomatically warned to keep away. The official Canadian view was that they did not want drilling to accidentally set off oil pollution at one or two promising sites. In Peterson's opinion, the Canadians did not want anyone besides themselves discovering oil in their seas and so, quite suddenly, their government announced that no ship like the *Challenger* could come within 200 miles of their shores.

Now the committee planning the *Challenger*'s trips

has to ask "Can we drill here? Who has the oil rights?" The NSF was also showing signs of setting boundaries on the exploration of the seas, in the name of efficiency and cost saving. Peterson was concerned that one day they might find Scripps ships could not go above 43° north latitude, for instance, because this territory now belonged exclusively to the University of Washington.

As Operations Manager of the Project, Ken Brunot loomed very large. Although not really tall, he gave the impression of being large physically and had a strong personality. Exactly the kind of man needed to deal with tough men in the crews of roughnecks. Yet Ken Brunot was a man with a very intellectual bent.

He had one language quirk. It was the constant use of the word "indeed." He would say, for instance, "It is *indeed* interesting." He also employed quite skillfully the language of the frontier, the southwest, and the oil business. On one occasion he said, "It wasn't one of your whiskey-drinking, kissing-in-the-mouth parties."

Brunot had studied geology at the University of California and hoped to go on to Scripps for a graduate degree in oceanography, but at the time he simply could not afford it. Therefore he went into the business of oil drilling, as his father had before him. He remained actively interested in the sciences of the seas, however, and when the DSDP was being organized his drilling experience plus his awareness of the progress in ocean studies made it quite natural for him to be asked to join up.

While working in the oil business, he was given a deep-think assignment. "Brunot," he was told, "I want you to consider in the broadest possible way how we can drill a better hole in the ground." Brunot's reaction was that the question should have been phrased in a different way. "How can we best extract hydrocarbons from the earth?" This seemed to him broader—a statement of the real problem.

He has at times taken this approach to deep-sea drilling. "Are we attacking this right? We're taking this land development [oil well drilling] to sea and trying to duplicate conditions by using stabilizers, dynamic positioning, and so forth. Maybe we need a new approach. Perhaps we could use satellites to give information to an inertial navigation system aboard ship. The computer on board ship would work to maintain the spot we want to be in, until the next time it could get a satellite navigational fix. This might be better than dropping expensive beacons to the bottom that can, as they have, malfunction.

"Maybe we should begin by working from the bottom. Or somewhere near it. Maybe a deep sea hovercraft or a Flip [a curious oceanographic vessel impervious to waves] that can ignore the heave that a surface ship gets. Or perhaps we could drill into the sea floor by means of a laser. The heat might turn the walls of the hole to glass and make a nice strong support. At desired levels in such a hole you could drill laterally to get your fossils."

16

"The earth sciences have evolved from the empirical and become global in nature. We now have access to the three-quarters of the globe that once were hidden. This has been attained, in part, because of the freedom of the high seas. But now constraints are showing up in that freedom of the seas. We can only hope that freedom for drilling does not come into jeopardy."

—MELVIN N. A. PETERSON,
Chief Scientist of the Deep Sea
Drilling Project

WHEN the *Challenger* set sail from Boston on Leg XII, it had already drilled for a total of 6,500 hours and traveled a distance equivalent to two trips around the world. The sites it had visited were mostly in the temperate or even tropical regions in order to find good weather but on Leg XII it ventured to latitude 60° North, up around Labrador, and the weather was cold, rainy, and foggy. Once on this trip the ship felt winds of 35 knots and swells 15 feet high. The scientists in charge were Dr. Anthony S. Laughton of England's National Institute of Oceanography in Surrey, and Dr. William A. Berggren of Woods Hole. Laughton, who received his graduate degree from Cambridge, had worked at Woods Hole in 1954 and 1955 and taken three cruises while there to study the sea floor. He had also done research work on the British oceanographic ship, *Discovery II*, and concentrated on geological studies of the Gulf of Aden, the Red Sea, and tiny Rockall Island in the North Atlantic west of Scotland.

Berggren had acted as a paleontologist on Leg I of the *Challenger* cruises, received his doctorate from the University of Stockholm, and worked for the Oasis Oil Company of Libya, in Tripoli. Also on board for this voyage were scientists from Canada, Germany, and Denmark.

One object of this far-north expedition was to discover if the Atlantic floor in the region had a different spreading rate or the same as that determined for the sea farther south. Some earlier data suggested that the spreading slowed down considerably, or even stopped, between 35 and 10 million years ago.

Cores taken above Labrador showed that the first Ice Age began three million years ago, rather than the one or two million years ago that had been the accepted figure before. Some sediments in the Labrador Sea turned out to be subtropical, indicating that a branch of the Gulf Stream had flowed along this coast before the Ice Ages started. Drilling in this region also proved that Canada and Greenland broke apart more than 100 million years ago.

After taking cores as it made its way across the Atlantic, the *Challenger* and its crew reached little Rockall Island, 250 miles northwest of Ireland and 225 miles west of the Hebrides. This isolated little cone, 100 yards in circumference and 70 feet high, is made of granite and is therefore continental. It rises above a sandbank 50 miles long and 25 miles wide. Rockall is a great place for fishing cod.

The scientific question was about the date when Rockall separated from Greenland, and from Europe. It was determined the separation from Greenland began 60 million years ago. Fifteen million years ago, Rockall started to sink, subside into the sea, and in 5 million years it sank to a depth of 5,000 feet, leaving above water only the fragment that exists as a menace to navigation today. At a press conference in London a reporter for the *Daily Telegraph* asked about Atlantis—a question that is a favorite of newspapermen. In reply Dr. Laughton said, "There was no sign of Atlantis in the Atlantic Ocean, as the continents sank about 50 to 60 million years ago, long before man appeared on this planet."

The Mid-Atlantic Ridge and its rift were not drilled into because the sediments there are so thin that the drill could not get the solid hold that it needed to core. The Bay of Biscay between France and Spain, however, was drilled in spite of the Spanish government, which was not enthusiastic about such geologizing in its waters. The *Challenger* stayed well offshore. The idea was to find out if the area originated because of the rotation of the Iberian peninsula (Spain and Portugal) south and east to its present position. The drilling seemed to indicate that Iberia had indeed drifted to its present position where it almost touches Africa at the Straits of Gibraltar. A fast rate of sediment deposition, from materials that originated on land, was associated with the active mountain building of the Alps 12 to 10 million years ago.

Docking at Lisbon in the middle of August, 1970, the *Challenger* once again changed scientists and crew. This time the leading scientists were Dr. William B. F. Ryan of Lamont and Dr. Kenneth Hsu, an American, but at the time teaching at the Swiss Federal Institute of Technology in Zurich.

Ryan went to Williams College, worked at Woods Hole, and went on many scientific cruises, then received his doctorate from Columbia University after writing a thesis on the geology of the Mediterranean. Although still quite young at the time of Leg XIII, Bill Ryan, some years ago at a conference of geologists in Argentina, spoke of the idea of continental drift very favorably. One of the strongest opponents of the drift idea, Dr. V. V. Beloussov of Russia, was in the audience. When Ryan concluded, Dr. Beloussov rose to ask him, in English, "Have you ever studied geology, sonny?"

Ryan worked very hard on the plans for Leg XIII. He believed that the Europeans are very "pro-Alps," certain their mountains are the source of all other geological events. To explain the various overlays of sediments that had already been found on the Mediterranean ocean floor, the site of his cruise, the Europeans wanted the "Med" to go up and down, like an elevator.

Ryan was very anxious that Europeans get involved in the DSDP and, except for Dr. Hsu, he was the only American scientist on the ship for this voyage. "One of our holes was planned by a French student on the western Mediterranean. If he can't sail with us from Lisbon, he'll come aboard by rowboat from Marseilles." There were eighty applications for the nine scientific positions aboard the ship. The most competent people to handle the expected problems were the ones chosen. The planners of the sites (and Ryan was anxious that they receive very full credit for their ideas) were not necessarily the best people to go to sea.

A concern for the *Challenger* people on this trip was the current political situation in the Mediterranean. Israel and the Arab states were at war, or a good facsimile of it. In addition, there were many salt domes (diapirs) in the eastern Mediterranean that might turn out to be full of oil and, thus, a potential for oil pollution existed. The State Department was quite aware of this *Challenger* voyage, and Ryan expected a radio message any moment ordering the scientists to call off the trip. The ship was near Greece the day President Nasser of Egypt died. Several months before this Mediterranean trip, Ryan believed that the United States government, although it might have wanted to do so, would not cancel this trip because such an act would alarm all the Europeans who had been invited to participate and a cancellation would naturally get into European newspapers. As it was, the *Challenger* did receive a visit from a United States destroyer in the Mediterranean, far out at sea. They were asked what the ship was and what it was doing there. The *Challenger*'s listening gear also received strange signals in the deep water, sounds that presumably came from the propellers of submarines attached to the U. S. Sixth (Mediterranean) Fleet.

Bill Ryan of Lamont does not want to be labelled as a Mediterranean man, although he is conscious of what an honor it is to be placed in charge of one of the cruises at his relatively young age.

The Co-Chief Scientist, Ken Hsu, who was born in Nanking, China, but is now a naturalized United States citizen, received his doctorate from the University of California at Los Angeles. He worked for the Shell Oil Company and several American Universities before taking a teaching position in Switzerland. Dr. Hsu was known as one of the oceanographers who still resisted the idea of continental drift. He had, therefore, been invited to go along on Leg III, the trip from Dakar to Rio across the South Atlantic. by Dr. Art Maxwell. The drilling results that Dr. Hsu observed on Leg III converted him to the new doctrine.

Leg XIII had Swiss, Italian, English, Austrian, and Roumanian people aboard. Maria Cita, from Milan, who seems to have been the Sophia Loren of this trip, as she had been on a previous *Challenger* voyage, was among those Europeans who joined the vessel in Lisbon.

For this leg, as Doctors Hsu and Ryan said on paper, "The main objective is to attempt to resolve questions on the origin of the Mediterranean. Many geologists think that the western Mediterranean basin was a part of the European continent, which was foundered some 20 to 30 million years ago. An alternative hypothesis suggests that Sardinia and Corsica were split away from Spain and that peninsular Italy separated from the main continental mass to the north by a counterclockwise rotation, leaving deep basins in their wake.

"The eastern Mediterranean may have a quite different history. We share a popular belief that this part of the Mediterranean may represent a relic of an extensive seaway between Eurasian and Indo-African continents. [Geologists usually call this old ocean the Tethys Sea.] Spreading of the floors of the Atlantic and Indian Ocean led to a collision of those continents, giving rise to a chain of mountains, extending from the Alps to the Himalayas. Only the eastern Mediterranean remains as an inland sea. If this view is correct, we may find some very ancient oceanic sediments.

"In summary, we expect the results of our Mediterranean findings would contribute greatly to our understanding of the mountain building process."

Drilling first of all in the Atlantic, southwest of Portugal, the scientists found sediments that were "strikingly similar" to the Jurassic sediments found on Leg XI in the western Atlantic. This great age suggested that Iberia was a "micro-continent" and broke off, with Africa, from Europe and North America more than 130 million years ago.

The weather on the trip was beautiful, except for one

mistral, a cold, strong wind out of the Alps, that reminded the men and women they were actually at sea on a ship (no matter how stabilized, how anti-roll). In the western Mediterranean they found evidence that the sea floor was profoundly affected by the relative drift of Europe and Africa away from North America. An excellent fit of sediments taken west of Corsica related almost exactly to sediments found on a steep cliff, an escarpment, off the south coast of France.

According to plate tectonic theory, Africa and Europe are crashing against each other. This motion has already created the Alps, and a great rise and deep sea trench system south of Greece is creating more mountains that will eventually, after North Africa has squeezed Greece and Turkey sufficiently, create more mountains in what is really the same chain. *Challenger* drilling confirmed this idea.

Drilling in all parts of the Mediterranean showed a surprising event that evidently occurred everywhere at the same time. At a certain depth, very salty evaporites were found, indications that the sea, or ocean, had once been very shallow and that fast evaporation (38 inches a year at the present time) had taken place. Some black layers in the sediments had mud cracks and they were so carboniferous (the precursor of oil or coal) that they would actually burn. The Mediterranean was once almost empty. If you had stood on a mountain high in Capri or outside Palermo you might have seen as flat a desert as the Mojave. This drying up was probably caused by the closing of the Straits of Gibraltar due to the motions of the great plates. Spain must have been touching North Africa and preventing Atlantic water from coming in. At present, the water in the Mediterranean evaporates so fast that it would dry up again if the Atlantic were not continually replenishing it. Between 12½ and 5 million years ago, it became so dry that fresh water lakes formed in some of the sea floor depressions. Now it is believed there is such an interchange of water between the Atlantic and the Mediterranean—the salty Mediterranean water flowing at the bottom of the sill and the fresher Atlantic water coming in at the surface—that erosion, no matter what the geological activity, will keep the Straits of Gibraltar open.

Carefully avoiding all the fascinating possible oil formations off the north coast of Africa and taking care not to get involved in the recurrent Mid-East political crisis, the *Challenger* returned to Lisbon. The cores will be analyzed for years but, meanwhile, the ship had to continue its grand program.

Dr. Dennis Hayes of Lamont and Dr. Tony Pimm, editor of several official *Challenger* books, were the scientists in charge as the vessel sailed into the Atlantic once more. In San Juan, Puerto Rico, they turned over the ship to Terry Edgar and John B. Saunders, a scientist from Trinidad. Terry Edgar had written his doctor's thesis on the nature of the Caribbean and was anxious to learn much more about this extremely complicated body of water. For instance, is the Puerto Rico Trench a very old feature that once lay in what is now the Pacific? Has it now become the Caribbean because the Americas have overtaken it and moved west? It was on Dr. Edgar's trip that the reentry system for drilling was first used. Then the ship was to go through the Panama Canal to investigate the Pacific ocean floor south of the Aleutian Islands and east of Japan to see if those areas were the oldest part of the oceans of the world. Then the *Challenger* would sail down to Australia, into the Indian Ocean, around South Africa, and eventually back to the ship's home port in Galveston, Texas. There are a considerable number of objectives on this great voyage. They will not all be realized, of course, and the findings will probably raise just as many questions as those that they answer.

As the *Challenger* was about to enter the Mediterranean, Dr. Melvin N. A. Peterson made an appearance before a U. S. Senate committee. He was recommending, without exactly saying so at the time, an extension of the program after the present agreement with the NSF had ended. After discussing some of the more spectacular achievements he said, in part, "We are in a period of profound change in our thinking about our planet and our relationship to it. If we take one example from Deep-Sea Drilling, to show the rapidity with which our concepts have changed, only a decade ago relatively few geologists espoused the idea of continental drift. At the present time, largely as the result of the interaction of oceanic geophysical studies and the results of Deep-Sea Drilling, the concepts of migrating continents, relative youth of the ocean basins, and continual formation of new oceanic crust along the axes of the Mid-Oceanic Rise-Rift systems, are virtually working tools of students of the earth.

"We are getting a first glimpse of the earth beneath the sea. I regard it as entirely within the role of responsible government, much in the manner of the Lewis and Clark expeditions, to make these first important observations. I would say we are in the 'Lewis and Clark stage' right now, as regards the ocean floor."

The epic voyages of the *Glomar Challenger* have already far surpassed, in distance at least, those of Lewis and Clark. Even if they were to end with an accident, or congressional apathy at the end of the present contract, the men and women who have worked aboard it will have changed for all time our view of the planet we must call our home.

SUGGESTED READING

BASCOM, WILLARD. *A Hole in the Bottom of the Sea.* New York: Doubleday & Co., 1962.

BRIGGS, PETER. *The Great Global Rift.* New York: Weybright and Talley, 1968.

GASKELL, BRUCE C., THARPE, MARIE, and EWING, MAURICE. *The Floors of the Oceans.* New York: The Geological Society of America, 1959.

HILL, MAURICE N., ed. *The Sea.* 3 vols. New York: John Wiley & Sons, 1963.

Initial Reports of the Deep Sea Drilling Project. Washington, D. C.: Superintendent of Documents, U. S. Government Printing Office. At the end of each leg of the *Glomar Challenger* cruises, the scientists aboard prepare a first, scholarly statement of their findings. These are published as soon as possible and each volume is catalogued under the name of the Co-Chief Scientists. They are edited for the National Science Foundation by the Scripps Institution of Oceanography of the University of California.

PICCARD, JACQUES, and DIETZ, ROBERT S. *Seven Miles Down: The Story of the Bathyscaph "Trieste."* New York: G. P. Putnam's Sons, 1950.

RAITT, HELEN. *Exploring the Deep Pacific.* New York: W. W. Norton & Co., 1964.

SHEPHARD, FRANCIS P. *Submarine Geology.* 2d ed. New York: Harper & Row, 1963.

Except for the *Initial Reports,* the story of the *Glomar Challenger* has not yet appeared in any books. However, articles about the ship, its findings and the whole question of continental drift and sea-floor spreading appear with great regularity in certain periodicals. *Scientific American, Science, Geotimes,* and *Journal of Geo-*

physical Research are the best sources for continuing information on this subject. They are indexed annually. Others are *Sea Frontiers, Ocean Industry, Bulletin of the American Association of Petroleum Geologists, Bulletin of the Geological Society of America, Deep Sea Research, Journal of Marine Research, Nature* (London), *The New Scientist* (London), and *Micropaleontology.*

Index

DATE DUE